Cambridge E

Elements in the Philoso|
edited by
Grant Ramsey
KU Leuven
Michael Ruse
Florida State University

GAMES IN THE PHILOSOPHY OF BIOLOGY

Cailin O'Connor
University of California, Irvine

CAMBRIDGE
UNIVERSITY PRESS

CAMBRIDGE
UNIVERSITY PRESS

University Printing House, Cambridge CB2 8BS, United Kingdom

One Liberty Plaza, 20th Floor, New York, NY 10006, USA

477 Williamstown Road, Port Melbourne, VIC 3207, Australia

314–321, 3rd Floor, Plot 3, Splendor Forum, Jasola District Centre, New Delhi – 110025, India

79 Anson Road, #06–04/06, Singapore 079906

Cambridge University Press is part of the University of Cambridge.

It furthers the University's mission by disseminating knowledge in the pursuit of education, learning, and research at the highest international levels of excellence.

www.cambridge.org
Information on this title: www.cambridge.org/9781108727518
DOI: 10.1017/9781108616737

First published 2020

A catalogue record for this publication is available from the British Library.

ISBN 978-1-108-72751-8 Paperback
ISSN 2515-1126 (online)
ISSN 2515-1118 (print)

Games in the Philosophy of Biology

Elements in the Philosophy of Biology

DOI: 10.1017/9781108616737
First published online: January 2020

Cailin O'Connor
University of California, Irvine

Abstract: This Element surveys the most important literature using game theory and evolutionary game theory to shed light on questions in the philosophy of biology. I focus on two branches of literature. I begin the Element with a short introduction to game theory and evolutionary game theory. After that, I turn to work using signaling games to explore questions related to communication, meaning, language, and reference. The second part of the Element addresses prosociality – strategic behavior that contributes to the successful functioning of social groups – using the prisoner's dilemma, stag hunt, and bargaining games.

Keywords: game theory, philosophy of biology, evolutionary theory, signaling, prosociality.

ISBNs: 9781108727518 (PB) 9781108616737 (OC)
ISSNs: 2515-1126 (online) 2515-1118 (print)

Contents

1 Introduction

A male peacock mantis shrimp resides happily in his burrow. He has recently molted, leaving his shell soft and vulnerable. As he waits, another male wanders onto his territory and approaches. The new male would like a burrow of his own. Both possess raptorial claws powerful enough to smash through the glass of an aquarium. If they fight for the territory, the temporarily squishy burrow owner will be seriously hurt. Neither, though, can directly observe whether the other has molted recently. Both mantis shrimp raise their appendages to display brightly colored "meral spots", intended to signal their aggression and strength. The intruder is impressed and backs off to seek territory elsewhere.

Alex is looking to hire a new data analyst for his company. Annaleigh wants the job and so is keen to impress Alex. She includes every accomplishment she can think of on her resume. Since she went to a prestigious college, she makes sure her educational background is front and center, where Alex will see it first.

A group of vampire bats need to eat nightly to maintain their strength. This is not always easy, however. Sometimes a bat will hunt all night but fail to find a meal. For this reason, most bats in the group have established relationships for reciprocal food sharing. Bats who manage to feed will regurgitate blood for partners who did not.

Mitzi needs a kidney transplant but is too old to go on the normal donor list. Her close friend wants to donate but is incompatible as a donor for Mitzi. They join the National Kidney Registry and become part of the longest-ever kidney donation chain. This involves thirty-four donors who pay forward kidneys to unrelated recipients so that their loved ones receive donations from others in the chain.[1]

We have just seen four examples of strategic scenarios. By strategic, I mean situations where (1) multiple actors are involved and (2) each actor is affected by what the others do. Two of these strategic scenarios I just described – the ones involving mantis shrimp and vampire bats – occurred between non-human animals. The other two – with Alex, Annaleigh, Mitzi, and her friend – occurred in the human social scene. Notice that strategic scenarios are completely ubiquitous in both realms. Whenever predators hunt prey, potential mates choose partners, social animals care for each other, mutualists exchange reciprocal goods or services, parents feed their young, or rivals battle over territory, these animals are engaged in strategic scenarios. Whenever humans do any of these things, or when they apply for jobs, work together to overthrow a dangerous authoritarian leader, bargain over the price of beans, plan a military

---

[1] This one is a true story. See UW Health (2015).

attack, divide labor in a household, or row a boat, they are likewise in strategic scenarios.

This ubiquity explains the success of the branch of mathematics called *game theory*. Game theory was first developed to explore strategic scenarios among humans specifically.[2] Before long, though, it became clear that this framework could be applied to the biological world as well. There have been some differences to how game theory has typically been used in the human realm versus the biological one. Standard game theoretic analyses start with a *game*. I will define games more precisely in Section 2, but for now it is enough to know that they are simplified representations of strategic scenarios. Game theoretic analyses typically proceed by making certain assumptions about behavior, most commonly that agents act in their own best interests and that they do this rationally.

Strong assumptions of rationality are not usually appropriate in the biological realm. When deciding whether to emit signals to quora of their peers, bacteria do not spend much time weighing the pros and cons of signaling or not signaling. They do not try to guess what other bacteria are thinking about and make a rational decision based on these calculations. But they do engage in strategic behavior, in the sense outlined above, and often very effectively.

There is another framework, closely related to game theory, designed to model just this sort of case. *Evolutionary game theory* also starts with games but focuses less on rationality.[3] Instead, this branch of theorizing attempts to explain strategic behavior in the light of evolution. Typical work of this sort applies what are called *dynamics* to populations playing games. Dynamics are rules for how strategic behavior will change over time. In particular, dynamics often represent evolution by natural selection. These models ask, in a group of actors engaged in strategic interactions, which sorts of behaviors will improve fitness? Which will evolve?

This is an Element about games in the philosophy of biology. The main goal here is to survey the most important literature using game theory and evolutionary game theory to shed light on questions in this field. Philosophy of biology is a subfield in philosophy of science. Philosophers of science do a range of things. Some philosophy of science asks questions like, How do scientists create knowledge? What are theories? And how do we shape an ideal science? Other philosophers of science do work that is continuous with the theoretical

---

[2] Von Neumann and Morgenstern (1944) originated the theory. Other very influential early contributions were made by John Nash (1950).

[3] This framework was first developed by Maynard-Smith and Price (1973). Before this, though, game theorists were already starting to think about dynamical approaches to games, as in Robinson (1951) and Brown (1951).

sciences, though often with a more philosophical bent. This Element will focus on work in this second vein.

In surveying this literature, then, it will not be appropriate to draw hard disciplinary boundaries. Some work by biologists and economists has greatly contributed to debates in philosophy of biology. Some work by philosophers of biology has greatly contributed to debates in the behavioral sciences. Instead, I will focus on areas of work that have been especially prominent in philosophy of biology and where the greatest contributions have been made by philosophers.

As I've also been hinting at, it will also not be appropriate to draw clear boundaries between biology and the social sciences in surveying this work. This is in part because game theory and evolutionary game theory have been adopted throughout the behavioral sciences – human and biological. Frameworks and results have been passed back and forth between disciplines. Work on the same model may tell us both about the behaviors of job seekers with college degrees and the behaviors of superb birds of paradise seeking mates, meaning that one theorist may contribute to the human and biological sciences at the same time.

There is something special that philosophers of biology have tended to contribute to this literature. Those trained in philosophy of science tend to think about science from a meta-perspective. As a result, many philosophers of science have both used formal tools and critiqued them at the same time. This can be a powerful combination. It is difficult to effectively critique scientific practice without being deeply immersed in that practice, but scientists are not always trained to turn a skeptical eye on the tools they use. This puts philosophers of biology who actually build behavioral models into a special position. As I will argue, this is a position that any modeler should ultimately assume – one of using modeling tools and continually assessing the tools themselves. Throughout this Element we will see how philosophers who adopted this sort of skeptical eye have helped improve the methodological practices of game theory and evolutionary game theory.

So, what are the particular debates and areas of work that will be surveyed here? I will focus on two branches of literature. I begin this Element with a short introduction to game theory and evolutionary game theory. After that, I turn to the first area of work, which uses *signaling games* to explore questions related to communication, meaning, language, and reference. There are three sections in this part of the Element. The first, Section 3, starts with the common interest signaling game described by David Lewis. This section shows how philosophers of science have greatly expanded theoretical knowledge of this model, while simultaneously using it to explore questions related to human and animal signaling. In Section 4, I turn to a variation on this model – the conflict of interest signaling game. This game models scenarios where actors communicate but do not always seek the same ends. As we will see, traditional approaches to conflict of interest signaling, which appeal to signal costs, have been criticized as biologically unrealistic and methodologically unsound. Philosophers of biology have helped to rework this framework, making clearer how costs can and cannot play a role in supporting conflict of interest signaling. The last section on signaling, Section 5, turns to deeper philosophical debates. I look at work using signaling games to consider what sort of information exists in biological signals. Do these signals include semantic content? Of what sort? Can they be deceptive?

The second part of the Element addresses a topic that has been very widely studied in game theory and evolutionary game theory – prosociality. By prosociality, I mean strategic behavior that generally contributes to the successful functioning of social groups. Section 6 looks at the evolution of altruism in the prisoner's dilemma game. This topic has been explored in biology, economics, philosophy, and all the rest of the social sciences, so this section is even more interdisciplinary than the rest of the Element. Because this game *has* been so widely studied, and the literature so often surveyed, I will keep this treatment very brief. Section 7 turns to two models that have gotten quite a lot of attention in philosophy of biology: the stag hunt and the Nash demand game. The stag hunt represents situations where cooperation is risky but mutually beneficial. The Nash demand game can represent bargaining and division of resources more generally. As we will see, these games have been used to elucidate the emergence of human conventions and norms governing things like the social contract, fairness, and unfairness.

By way of concluding the Element, I'll dig a bit deeper into a topic mentioned above: methodology. In particular, a number of philosophers have made significant methodological contributions to evolutionary game theory that stand free of particular problems and applications. The epilogue, Section 8, briefly discusses this literature.

Readers should note that the order and focus of this Element are slightly idiosyncratic. A more traditional overview might start with the earliest work in evolutionary game theory – looking at the hawk-dove game in biology – and proceed to the enormous literature on cooperation and altruism that developed after this. The literature on signaling games would come later, reflecting the fact that it is relatively recent, and would take up a less significant chunk of the Element. My goal is to give a more thorough overview of work that has received less attention rather than to revisit well-worn territory. In addition, as mentioned, I focus in this Element on the areas of the literature that have been most heavily developed in philosophy.

This Element is very short. For this reason, it does not provide a deep understanding of any one topic but overviews many. Interested readers should, of course, use the references in this Element to dig deeper. Furthermore, the brevity of the Element means there is no space to even briefly touch on a number of related literatures. Most notably, I ignore important work in biology and philosophy of biology on other sorts of evolutionary theory/modeling and work in economics on strategic behavior in the human realm.

So, in the interest of keeping it snappy, on to Section 2, and the introduction to game theory and evolutionary game theory.

2 Games and Dynamics

What does a strategic scenario involve? Let's use an example to flesh this out. Every Friday, Alice and Sharla like to have coffee at Peets. Upon arriving slightly late, though, Sharla notices it is closed and Alice is not there. Alice has likely chosen another coffee shop – either the nearby Starbucks or the Dunkin' Donuts. Which should Sharla try?[4] This sort of situation is often referred to as a coordination problem. Two actors would like to coordinate their action, in this case by choosing the same coffee shop. They care quite a lot about going to the same place, and less about which place it is.

How can we formalize this scenario into a game? We do this by specifying four things. First, we need to specify the *players*, or who is involved. In this case, there are two players, Alice and Sharla. If we like, we can call them player 1 and player 2. Second, we need to specify what are called *strategies*. Strategies capture the behavioral choices that players can make. In this case, each player has two choices, to go to Starbucks or to go to Dunkin' Donuts. Let's call these strategies *a* and *b*. Third, we need to specify what each player gets for different combinations of strategies, or what her *payoffs* are.

This is a bit trickier. In reality, the payoff to each player is the joy of meeting a friend should they choose the same strategy and unhappiness over missing each other if they fail. But to formalize this into a game, we need to express these payoffs mathematically. What we do is choose somewhat arbitrary numbers meant to capture the *utility* each player experiences for each possible outcome. Utility is a concept that tracks joy, pleasure, preference, or whatever it is that

---

[4] Let's suppose neither owns a cell phone.

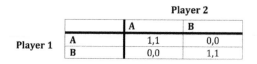

		Player 2	
		A	**B**
Player 1	**A**	1,1	0,0
	B	0,0	1,1

Figure 2.1 A payoff table of a simple coordination game. There are two players, each of whom chooses *a* or *b*. Payoffs are listed with player 1 first.

players act to obtain.[5] Let us say that when Sharla and Alice meet, they get payoffs of 1, and when they do not, they get lower payoffs of 0.

We can now specify a game representing the interaction between Alice and Sharla. Figure 2.1 shows what is called a *payoff table* for this game. Rows represent Alice's choices, and columns represent Sharla's. Each entry shows their payoffs for a combination of choices, with Alice's (player 1's) payoff first. As specified, when they both pick *a* or both pick *b*, they get 1. When they miscoordinate, they get 0. Game theorists call games like this – ones that represent coordination problems – *coordination games*. This game, in particular, we might call a pure coordination game. The only thing that matters to the actors from a payoff perspective is whether they coordinate.

There is one last aspect of a game that often needs to be specified, and that is *information*. Information characterizes what each player knows about the strategic situation – does she know the structure of the game? Does she know anything about the other player? In evolutionary models the information aspect of games is often downplayed, though. This is because information in this sense is most relevant to rational calculations of game theoretic behavior, not to how natural selection will operate.

So now we understand how to build a basic game to represent a strategic scenario. The next question is, how do we analyze this game? How do we use this model to gain knowledge about real-world strategic scenarios? In classic game theory, as mentioned in the introduction, it is standard to assume that each actor attempts to get the most payoff possible given the structure of the game and what the player knows. This assumption allows researchers to derive

---

[5] Utility is a controversial concept. It has been criticized as leading to circular reasoning – players act to get utility because utility is the sort of thing players act to get (Robinson, 1962). In evolutionary models, as we will see, this circularity is less of a worry. A traditional justification of the utility concept in economics stems from Von Neumann and Morgenstern (1944) (this justification was outlined in the second, 1947 edition of their book). They show that agents who satisfy four reasonable axioms have preferences that can be respresented by a utility function. And if these agents act to satisfy these preferences, they will act as if they are maximizing expected utility.

predictions about which strategies a player will choose, or might choose, and also to explain observed behavior in strategic scenarios.

More specifically, these predictions and explanations are derived using different solution concepts, the most important being the *Nash equilibrium* concept. A Nash equilibrium is a set of strategies where no player can change strategies and get a higher payoff. There are two Nash equilibria in the coordination game – both players choose *a* or both choose *b*.[6] Although this is far from the only solution concept in use by game theorists, it is most used by philosophers of biology, who tend to focus on evolutionary models. As we will see, Nash equilibria have predictive power when it comes to evolution, as well as to rational choice–based analyses.[7]

Notice that there is something in the original scenario with Sharla and Alice that is not captured in the game depicted in Figure 2.1. Remember that we supposed Sharla arrived at Peets late, to discover Alice had already made a choice. The payoff table above displays what is called the *normal form* coordination game. Normal form games do not capture the fact that strategic decisions happen over time. Sometimes one cannot appropriately model a strategic scenario without this time progression. In such cases, one can build a game in *extensive form*. Figure 2.2 shows this. Play starts at the top node. First Alice (player 1) chooses between Starbucks and Dunkin' Donuts (*a* or *b*). Then Sharla (player 2)

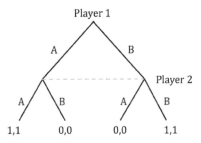

Figure 2.2 The extensive form of a simple coordination game. There are two players, each of whom chooses *a* or *b*. Player one chooses first.

---

[6] This is imprecise. In fact, there are only two *pure strategy* Nash equilibria. These are equilibria in strategies where agents always take the same behavior. At some points in the Element we will look at *mixed strategies*, where agents probabilistically mix behaviors. Many games have Nash equilibria in mixed strategies, and some games only have Nash equilibria in mixed strategies, though we will only discuss these when they are evolutionarily relevant.

[7] This predictive power should not be taken too strongly. Experimental evidence finds that in some cases, laboratory subjects play Nash equilibria in the lab, and in other cases not (Smith, 1994). For instance, Güth et al. (1982) give a famous example of the failure of Nash equilibrium predictions to account for bargaining behavior.

chooses. The dotted line specifies an *information set* for Sharla. Basically it means that she does not know which of those two nodes she is at, since she did not observe Alice's choice. At the ends of the branches are the payoffs, again with player 1 listed first. There are ways to analyze extensive form games that are not possible in normal form games, but discussing these goes beyond the purview of this Element.

Now let us turn to evolutionary analyses. In evolutionary game theoretic models, groups of individuals are playing games, and evolving or learning behavior over time. *Dynamics* represent rules for how evolution or learning occurs. The question becomes: what behaviors will emerge over the course of an evolutionary process?

One central approach developed in biology to answer this question involves identifying what are called *evolutionarily stable strategies* (ESSs) of games. Intuitively, an ESS is a strategy that, if played by an entire population, is stable against mutation or invasion of other strategies.[8] Thus ESSs predict which stable evolutionary outcomes might emerge in an evolving population. Despite being used for evolutionary analyses, though, the ESS concept is actually a static rather than an explicitly dynamical one. Without specifying a particular dynamic, one can identify ESSs of a game.

Here is how an ESS is defined. Suppose we have strategies a and b, and let $u(a, b)$ refer to the payoff received for playing strategy a against strategy b. Strategy a is an ESS against strategy b whenever $u(a, a) > u(b, a)$, or else if $u(a, a) = u(b, a)$, then it is still an ESS if $u(a, b) > u(b, b)$. When these conditions hold, if an a type mutates into a b type, we should expect this new b type to die off because they get lower payoffs than as do. The strategies here are usually thought of as corresponding to fixed behavioral phenotypes because biological evolution is the usual target for ESS analysis. In the cultural realm, one can apply the ESS concept by thinking of mutation as corresponding to experimentation. The fit is arguably less good, but an ESS analysis can tell us what learned behaviors might die out because those who experiment with them will switch back to a more successful choice.

Let's clarify with an example. Suppose we have a population playing the game in Figure 2.1 – they all choose which coffee shop to meet at. Is a (all going to Starbucks) an ESS? The first condition holds whenever the payoff one gets for playing a against a is higher than the payoff for playing b against a. This is true. So we know a is an ESS by the first condition. Using the same reasoning, we can see that b (all going to Dunkin' Donuts) is an ESS too. Intuitively this

---

[8] This concept was introduced and defined by Maynard-Smith and Price (1973). Another related concept is the evolutionarily stable state, which need not involve only one strategy.

should make sense. When we have a population of people going to one coffee shop (who all like to coordinate) no one wants to switch to the other.

The second condition comes into play whenever the payoff of *a* against *a* is the same as the payoff of *b* against *a*. This would capture a scenario where a new, invading strategy does just as well against the current one as it does against itself. We can see that *a* will still be stable, though, if *b* does not do very well against itself.

As mentioned, Nash equilibria are important from an evolutionary standpoint. In particular, every ESS will be a Nash equilibrium (though the reverse is not true). So identifying Nash equilibria is the first step to finding ESSs.

As we will see in Section 8, philosophers of biology have sometimes been critical of the ESS concept. A central worry is that in many cases full dynamical analyses of evolutionary models reveal ESS analyses to be misleading. For example, some ESSs have very small *basins of attraction*. A basin of attraction for an evolutionary model specifies what proportion of population states will end up evolving to some equilibrium (more on this shortly). In this way, basins of attraction tell us something about how likely an equilibrium is to arise and just how stable it is to mutation and invasion. Another worry is that sometimes stable states evolve that are not ESSs, and are thus not predicted by ESS analyses.

We will return to these topics later. For now, let us discuss in more detail what a full dynamical analysis of an evolutionary game theoretic model might look like. First, one must make some assumptions about what sort of population is evolving. A typical assumption involves considering an uncountably infinite population. This may sound strange (there are no infinite populations of animals), but in many cases, infinite population models are good approximations to finite populations, and the math is easier. One can also consider finite populations of different sizes.

Second, one must specify the interactive structure of a population. The most common modeling assumption is that all individuals belong to a single population that freely mixes. This means that every individual meets every other individual with the same probability. This assumption, again, makes the math particularly easy, but it can be dropped. For instance, a modeler might need to look at models with multiple interacting populations. (This might capture the evolution of a mutualism between two species, for example.) Or it might be that individuals tend to choose partners with their own strategies. Or perhaps individuals are located in a spatial structure which increases their chances of meeting nearby actors.

Last, one must choose dynamics which, as mentioned, define how a population will change over time. The most widely used class of dynamics – *payoff monotonic dynamics* – makes variations on the following assumption: whatever strategies receive higher payoffs tend to become more prevalent, while strategies that receiver lower payoffs tend to die out. In particular, the *replicator dynamics* are the most commonly used dynamics in evolutionary game theory.[9] They assume that the degree to which a strategy over- or underperforms the average determines the degree to which it grows or shrinks. We can see how such an assumption might track evolution by natural selection – strategies that improve fitness lead to increased reproduction of offspring who tend to have the same strategies. This model can also represent cultural change via imitation of successful strategies and has thus been widely employed to model cultural evolution.[10] The assumption underlying this interpretation is that group members imitate strategies proportional to the success of these strategies. In this Element, we will mostly discuss work using the replicator dynamics, and other dynamics will be introduced as necessary.

Note that once we switch to a dynamical analysis, payoffs in the model no longer track utility, but instead track whatever it is that the dynamics correspond to. If we are looking at a biological interpretation, payoffs track fitness. If we are looking at cultural imitation, payoffs track whatever it is that causes imitation – perhaps material success.

Note also that evolutionary game theoretic dynamics do not usually explicitly model sexual reproduction. Instead, these models usually make what is called the "phenotypic gambit" by assuming asexually reproducing individuals whose offspring are direct behavioral clones. Again, this simplifies the math, and again, many such models are good enough approximations to provide information about real, sexually reproducing systems.

To get a better handle on all this, let us consider a population playing the coordination game. Assume it is an infinite population and that every individual freely mixes. Assume further that it updates via the replicator dynamics.

The standard replicator dynamics are *deterministic*, meaning that given a starting state of a population, they completely determine how it will evolve.[11] This determinism means that we can fully analyze this model by figuring out what each possible population state will evolve to. We can start this analysis by looking at the equation specified for this model and figuring out what

---

[9] These were introduced by Taylor and Jonker (1978) with the goal of providing a truly dynamic underpinning for ESS analyses.

[10] The replicator dynamics are the *mean field equations* of explicit models of cultural learning. This means that they represent the average, expected change in these stochastic models (Weibull, 1997).

[11] A *stochastic* dynamic on the other hand will incorporate some randomness so that, often, one population state has the potential to end up at different equilibria.

the *rest points* of the system are.[12] These are the population states where evolution stops, so they are the candidates for endpoints of the evolutionary process. In particular we can ask, which of these rest points are ones that other states evolve to? Isolating these will tell us which stable outcomes to expect. Note that these population rest points are themselves often referred to as equilibria, since these are balanced points where no change occurs. These are equilibria of the evolutionary model, not of the underlying game. This said, Nash equilibria – and thus ESSs – are always rest points for the replicator dynamics.[13]

In this case, there are three equilibria, which are the states where everyone plays *a*, everyone plays *b*, and where there are exactly 50 percent of each. This third equilibrium is not a stable one, though, in the sense that a population perturbed away from it even a bit will evolve toward one of the other two equilibria. This means that we have a prediction as to what will evolve – either a population where everyone plays *a* or everyone plays *b*. Notice that these are the two ESSs.

We can say more than this. As mentioned, we can also specify just which states will evolve to which of these two equilibria. For this model, the replicator equation tells us that whenever more than one-half of individuals play *a*, the *a* types will become more prevalent. Whenever less than one-half play *a*, the *b* types become more prevalent. This makes intuitive sense – those playing the less prevalent strategy coordinate less often, do worse, and become even less prevalent. We can summarize these results in a *phase diagram*, or a visual representation of each population state and where it evolves.

Figure 2.3 shows this. The line displays proportions of strategies from 0 percent *a* (or all *b*) to 100 percent *a*. The two endpoints are the two stable equilibria. The open dot in the middle represents the unstable equilibrium. Arrows show the direction of population change, toward these two equilibria. The diagram

---

[12] Let strategies be indexed by x_i. The replicator equations are

$$\dot{x}_i = x_i(f_i(x) - \sum_{j=1}^{n} f_j(x)x_j). \tag{2.1}$$

This says that the change in strategy i, ($\dot{x}_i$) is equal to the current proportion of this strategy, x_i, multiplied by the difference between the average payoff generated by i ($f_i(x)$) minus the average fitness for all strategies in the population ($\sum_{j=1}^{n} f_j(x)x_j$). For the coordination game, this yields one equation for the change in *a* (the change in *b* is 1 minus the change in *a*): $\dot{x} = 3x^2 - x - 2x^3$. Notice that $\dot{x} = 0$ when $x = 0, .5, 1$, which are the rest points. When $0 \leq x \leq .5$, this change will be negative, meaning that when the proportion of *a* is less than half, it decreases. When $.5 \leq x \leq 1$, this change is positive meaning that the proportion of *a* is increasing.

[13] This does not mean that all Nash equilibria correspond to stable, or evolutionarily significant rest points. Also note that the reverse is not true – there are rest points of the replicator dynamics that are not Nash equilibria. For more, see Sigmund (2011).

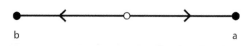

Figure 2.3 The phase diagram for a population evolving via the replicator
dynamics to play a simple coordination game.

also illustrates the *basins of attraction* for each equilibrium, or the set of starting points that will end up at that equilibrium. The points to the right of the unstable rest point are the basin of attraction for the *a* equilibrium, and the points to the left for the *b*. In this case, the two basins of attraction have equal size, but, as we will see, in many models they differ. The size of a basin of attraction is often taken to give information about the likelihood that an equilibrium evolves.

This is a very simple evolutionary model. Models with more strategies and more complicated population structures will be harder to analyze. A host of methods can be employed to learn about these evolutionary systems. For example, one might run simulations of the replicator dynamics for thousands of randomly generated population states. Doing so generates an approximation of the sizes of the various basins of attraction for a model. One might also use analytic methods to prove various things about the evolutionary system – for example, that all population states will evolve to some particular rest point – even if one cannot generate a complete dynamical picture.

At this point, we have surveyed some general concepts in game theory and evolutionary game theory. We have also seen an example of how one might analyze a simple game theoretic and evolutionary game theoretic model. This, of course, is a completely rudimentary introduction, but I hope it will suffice for readers who would like to situate the rest of this Element. Those who are interested in learning more can look at Weibull (1997) or Gintis (2000). As we will see, philosophers of biology have made diverse uses of these frameworks, contributing to diverse literatures. Let us now turn to the first of these literatures – that on signaling.

3 Common Interest Signaling

Baby chicks have a call that they issue when lost or separated from their mother. This induces the mother chicken to rush over and reclaim the chick. This call is an example of a *signal*, an arbitrary sign, symbol, or other observable, that plays a role in communication between organisms. By arbitrary, I simply mean that the signal could play its role even if it was something else. For the chickens, many different calls could have done just as well. We see such arbitrary signals throughout the human social world, and across the rest of the animal kingdom (and even outside it). Humans communicate with each other using language, both spoken and written, and a bestiary of other signals, including hand gestures, signage, facial expressions, etc. Many other species engage in verbal signaling. A famous example comes from vervet monkeys who have a host of evolved alarm calls (Seyfarth et al., 1980b). Other animals, like wolves, signal with stereotyped gestures and body positions (Zimen, 1981). Some animals send chemical signals, including pheromones. In addition, for many animals, actual features of their physiology serve as signals. A classic example is the tail of the peacock, though we will focus more on this sort of case in Section 4, when we talk about conflict of interest signaling.

There is a reason for the ubiquity of signaling. It can be incredibly beneficial from a payoff standpoint for organisms to communicate about the world around them, and about themselves. At the most basic biological level, we would not be here without genetic signaling. As Skyrms (2010b) puts it, "signals transmit information, and it is the flow of information that makes all life possible" (32).

Of course, the actual details of what a signal is, who is communicating, and for what purpose vary across these examples. What unifies them is that in each case organisms derived enough benefit from communicating that they evolved, or culturally evolved, or learned signaling conventions to facilitate this communication. As will become clear, we can unify these diverse scenarios under one model, which helps us understand all sorts of things about them. The goal of this section will be to introduce this framework and describe some ways philosophers have used it to inform questions in philosophy, biology, and the social sciences.

I will start by describing the common interest signaling game and briefly discussing its history in philosophical discourse. After that, I'll describe how philosophers have been instrumental in developing our theoretical understanding of the evolution of signaling using this model. I will then move on to describe some of the diverse ways that the signaling model has been applied, including to human language, animal communication, and communication within the body.

3.1 The Lewis Signaling Game

Quine ([1936] 1976) famously argued that linguistic conventions could not emerge naturally. The idea is that one cannot assign meaning to a term without using an established language. In *Convention*, David Lewis first employs game theory to chip away at this argument (Lewis, 1969). In particular, he introduces

a highly simplified game to show how terms might attain conventional meaning in a communication context. Philosophers have often called this the *Lewis
signaling game*, though it is also referred to as the *common interest signaling
game*, or just the signaling game. As we will see, there are actually a large
family of games that go under this heading, but let us start with the basics.

Suppose that a mother and a daughter regularly go on outings. Sometimes
it is rainy, in which case they like to walk in the rain. If it is sunny, they have
a picnic outside. To figure out whether to bring umbrellas or sandwiches, the
daughter calls her mother ahead of time to see what the weather is like. The
mother reports that it is either "sunny" or "rainy," and the daughter makes her
choice by bringing either umbrellas or a picnic.

We might represent a scenario like this with the following game. There are
two possible states of the world – call them state 1 and state 2 – which represent
the weather being sunny or rainy. A player called the *sender* (the mother) is able
to observe these states and contingent on that observation to send one of two
signals ("sunny" or "rainy" – call them signal a and signal b). A player called the
receiver, the daughter, cannot observe the world directly. Instead, she receives
whatever message is sent and can take one of two actions (act 1 and act 2) –
bringing umbrellas or a picnic. Act 1 is appropriate in state 1 and act 2 in state
2. If the receiver takes the appropriate action, both players get a payoff of 1 (just
like the mother and daughter get a payoff for a picnic in the sun and umbrellas
in the rain), otherwise they get nothing. In this game, two players want the
same thing – for the action taken by the receiver to match the state observed by
the sender. (This is why the game is a *common interest* signaling game.) The
only way to achieve this coordination dependably is through the transmitted
signal.

Figure 3.1 shows the extensive form of this game. This game tree should be
read starting with the central node. Remember from the last section that play of
the game can go down any of the forking paths depending on what each agent
chooses to do. In order to capture the fact that one of two states obtains inde-
pendent of the players' choices, we introduce a dummy player called *Nature*.
Nature (N) first chooses state 1 or state 2. Then the sender (S) chooses signal
a or b. The receiver (R) has only two information sets since they do not know
which state obtains. These, remember, are the nodes connected by dotted lines.
The receiver, upon finding themselves at one of these sets, chooses act 1 or 2.
The payoffs are listed at the terminal branches. Since payoffs are identical for
both actors, we list only one number.

Lewis observed that there are two special sorts of Nash equilibria of this
game often called *signaling systems*. These are the sets of strategies where
the sender and receiver always manage to coordinate with the help of their

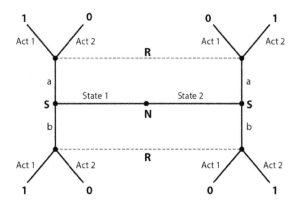

Figure 3.1 The extensive form respresentation of a Lewis signaling game with two states, two signals (a and b), and two acts.

State	Signal	Act	State	Signal	Act
1 → a → 1			1 ⟶ a ⟶ 1		
2 → b → 2			2 ⟶ b ⟶ 2		

Figure 3.2 The two signaling systems of the Lewis signaling game. In the first, signal a represents state 1 and causes act 1, while signal b attaches to state 2 and act 2. In the second system, the meanings of the two signals are reversed.

signals. Either signal a is always sent in state 1, and signal b in state 2, and the receiver always uses these to choose the right act, or else the signals are flipped. Figure 3.2 shows these two signaling systems. Arrows map states to signals and signals to acts. Notice that the left arrows in each diagram represent the sender's strategy, and the right ones the receiver's.

What Lewis argued is that at each of these equilibria, players engage in a linguistic *convention* – a useful, dependable social pattern which might have been otherwise. (In Section 7 we will say a bit more about this concept.) For Lewis, these conventions might arise through a process of precedent setting or as a result of natural salience. If so, this could explain how conventional meaning might emerge naturally in human groups.

But one might worry that while Lewis gestures toward a process by which signals can obtain conventional meaning, he does not do much to fill in how such a process might go. Furthermore, we have seen that conventional meaning has emerged throughout the natural world. Can Lewis's analysis address vervet monkeys? We do not suppose that their alarm signals were developed through a process of precedent setting between members of the group. Furthermore, Lewis requires that his players have full common knowledge of the game and expectations about the play of their partner for an equilibrium to count as a convention. These conditions will not apply to the animal world (and sometimes not to the human one either). Let us now turn to explicitly evolutionary models of this game.

3.2 The Evolution of Common Interest Signaling

Skyrms (1996) uses evolutionary game theory to strengthen Lewis's response to Quine. He starts by observing that in the simple signaling game described above, the only ESSs are the two signaling systems (Wärneryd, 1993).[14] This is because each signaling system achieves perfect coordination, and any other strategy will sometimes fail to coordinate with those playing the signaling system. In addition no other strategy has this property – each may be invaded by the signaling system strategies (Wärneryd, 1993; Skyrms, 1996).[15] So in an evolutionary model, then, we have some reason to think the signaling systems will evolve.

As Skyrms points out, true skeptics might not be moved by this ESS analysis. In particular, they might worry that there is no reason to think that either one signaling system or the other will take over a population. If so, groups might fail to evolve conventional meaning. To counter this worry, he analyzes what happens when a single, infinite population evolves to play the signaling game under the replicator dynamics. We observed in Section 2 that these dynamics can represent both cultural evolution and natural selection. If we see meaning emerge in this model, then, we have good reason to think that simple signals can emerge in the natural world as well as the social one. Using computer simulation tools, Skyrms finds that one of the signaling systems always takes over the population. Subsequently, proof-based methods have been used to show this will always happen (Skyrms and Pemantle, 2000; Huttegger, 2007a).[16] As Skryms concludes for this model, "the emergence of meaning is a moral certainty" (93).

As we will see in a minute, this statement must be taken with a grain of salt. But this "moral certainty" nonetheless has explanatory power. Signaling

---

[14] See also Blume et al. (1993), who consider more general games.

[15] To find ESSs here, one can symmetrize the game by allowing each actor to have both a sender and a receiver strategy. Or else one can analyze two populations, one of senders and the other receivers.

[16] And Skyrms and Pemantle (2000) and Huttegger (2007c) show the robustness of this evolution under different dynamics.

is ubiquitous across the biological and human social realms, and now we see part of the reason why. Despite Quine's worries, signaling is easy to evolve. It takes very little to generate conventional meaning from nothing.

This is not the end of the story, though. Subsequently, philosophers and others have developed a deep understanding of common interest signaling, deriving further results about this very simplest signaling model and a host of variants. In addition, the signaling game has been co-opted in numerous ways to address particular research questions. Let us first look at what else we have learned about the signaling game, and then move on to some applications. Much of the literature I will now describe is discussed in more detail in Skyrms (2010b).[17]

The simple game we have been discussing is sometimes called the $2 \times 2 \times 2$ game, for two states, two signals, and two possible actions. But we can consider $N \times N \times N$ games with arbitrary numbers of states, signals, and actions. When $N > 2$, these models have a new set of equilibria called *partial pooling equilibria*. The sender sends the same message in multiple states, and upon receipt of this partially pooled message the receiver mixes between the actions that might work. Figure 3.3 shows what this might look like for the $3 \times 3 \times 3$ game. After receipt of signal b, the receiver does act 2 with probability x and act 3 with probability $1 - x$. These are *mixed strategy equilibria*, where the actors mix over behaviors.[18] The only ESSs of these games are the signaling systems, not these partial pooling equilibria. These equilibria, however, do have basins of attraction under the replicator dynamics (Nowak and Krakauer, 1999; Huttegger, 2007a; Pawlowitsch et al., 2008). And for larger games, simulations show an increasing likelihood that a system of partial communication will

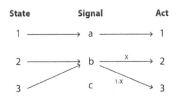

Figure 3.3 A partial pooling equilibrium in a $3 \times 3 \times 3$ signaling game. The sender uses b in both states 2 and 3. Upon receipt of b, the receiver mixes between acts 2 and 3.

---

[17] Huttegger and Zollman (2011b) and Huttegger et al. (2014) give more technical overviews.
[18] Notice that this means there is actually a continuum of equilibria for each value of x. Of course, there are also other continua where the sender pools states 1 and 2 or states 1 and 3, and uses different signals to do so.

emerge instead of perfect signaling (Barrett, 2009). In other words, the moral certainty of signaling is disrupted when even a bit of complexity is added to the model.

Another, even smaller change can disrupt this certainty. Consider the $2 \times 2 \times 2$ model where one state is more probable than the other. The more common one state is, the smaller the benefits of signaling. (When one state occurs 90 percent of the time, for example, the receiver can be very successful by simply taking the action that is best in that state.) As a result, *babbling equilibria*, where no information is transferred because sender and receiver always do the same thing, evolve under the replicator dynamics with greater likelihood the more likely one state is (Huttegger, 2007a; Skyrms, 2010b).

There is a methodological lesson here. Early analyses of ESSs in common interest signaling games, as in Wärneryd (1993), indicated that perfect signaling is expected to evolve. But, as we have just noted, more complete dynamical analyses show the ESS analysis insufficient – it fails to capture the relevance and stability of babbling equilibria and partial pooling equilibria.

There is also a philosophical lesson. The fact that these small changes lead to situations where perfect signaling does not emerge seems to press against the power of these models to help explain the ubiquity of real-world signaling. But we should not be too hasty. Philosophers have looked at changes to these more complex models that increase the probability of signaling. These include the addition of drift and mutation to the models (Huttegger, 2007a; Hofbauer and Huttegger, 2008), correlation between strategies – where signalers tend to meet those of their own type (Skyrms and Pemantle, 2000; Huttegger and Zollman, 2011a; Skyrms, 2010b), and finite populations (Pawlowitsch, 2007; Huttegger et al., 2010). Furthermore, even in cases where evolved communication is not perfect, quite a lot of information may be transferred (Barrett, 2009). In other words, our simple signaling models retain their power to help explain the ubiquity of real-world signaling, even if in many cases perfect signaling does not emerge.

Let's look at some more variants that extend our understanding of signaling. Sometimes more than two agents communicate at once. Bacterial signaling is usually between multiple senders and multiple receivers, for instance. And vervet monkeys send alarm calls to all nearby group members. Barrett (2007) and Skyrms (2009) consider a model with two senders, one receiver, and four states of the world, where senders each observe different partitions of the states. (That is, one sender can distinguish states 1–2 from states 3–4, and the other can distinguish odd and even states.) These authors show how signaling systems can emerge that perfectly convey information to the receiver.[19] Skyrms (2009, 2010b) considers further variations, such as models where one sender communicates with two receivers who must take complimentary roles, where a receiver listens to multiple senders who are only partially reliable, where actors transmit information in signaling "chains," and dialogue models where the receiver tells the sender what information they seek and the sender responds. In each case, the ease with which signaling can emerge is confirmed.

Some creatures may be limited in the number of signals they can send. Others (like humans) may be able to easily generate a surfeit of terms. In an evolutionary setting, games with too many signals often evolve to outcomes where one state is represented by multiple terms or "synonyms". In these states, actors coordinate just as well as they would with fewer signals. This is because each signal has one meaning, even if a state may be represented by multiple signals (Pawlowitsch et al., 2008; Skyrms, 2010b).

Without enough signals, actors cannot perfectly coordinate their action with the world. This is because they do not have the expressive power to represent each state individually. For this reason, Skyrms (2010b) describes this sort of situation as an "information bottleneck."[20] At the optimal equilibria of such models, actors may either ignore certain states or else categorize multiple states under the same signal. Donaldson et al. (2007) point out that in some cases actors have payoff reasons for choosing some ways of categorizing over others, in which case these ways may be more likely to evolve. For example, while the ideal response to predatory wolves and dogs might be slightly different for a group of monkeys, it still might make sense to use one alarm call for these highly similar predators if monkeys are limited in their number of signals. We will come back to this theme in Section 3.3.

One thing actors without enough signals might like to do is to invent new ones – this should help them improve coordination. To add signals, though, involves a change to the structure of the game itself. Skyrms (2010b) uses a different sort of model – one that explicitly represents learning – to explore this possibility. Such models have provided an alternative paradigm to explore the evolution of signaling, one that has often created synergy with evolutionary models. Let us now turn to this paradigm.

---

[19] Barrett (2007, 2010) uses the model to demonstrate how "kind terms" might themselves be conventional.

[20] Notice that what we designate as a state in the signaling game is arbitrary. For this reason, it makes sense to say there are more states than signals whenever actors do not have the language to differentiate states that they can gain a payoff benefit from differentiating. In other words, if two states can be responded to with the same act, they should be treated as one state for modeling purposes.

3.2.1 Learning, Simulation, and the Emergence of Signaling

The results discussed thus far are from models that assume infinite populations, and change via either natural selection or cultural imitation. There is another set of models that have been widely used by philosophers to investigate signaling that instead represent two agents – a sender and a receiver – who learn to signal over time. It should be clear why such a model is a particularly good one to respond to skeptical worries about conventional meaning. They show that not only can such meaning evolve over time, it can emerge on very short timescales between agents with simple learning rules.

This framework assumes two agents play the signaling game round after round and that over time, they learn via *reinforcement learning*. This is a type of learning that uses past successes and failures to weight the probabilities that actors use different strategies in future play. One widely used reinforcement learning rule was introduced by Roth and Erev (1995).[21] The general idea is that actors increase the likelihood that they take a certain action proportional to the success of that action in the past.

Argiento et al. (2009) prove that actors in the $2 \times 2 \times 2$ reinforcement learning model always learn signaling systems. In other words, the lesson about moral certainty transfers to the learning case. There is something especially nice about this result, because the learning dynamics are so simple. Any critter capable of performing successful actions with increasing likelihood will be able to learn to signal. Once again, models make clear that meaning is easy to get.

Many of the other results from evolutionary models transfer to these learning models too. Barrett (2009) and Argiento et al. (2009), for example, show that in $N \times N \times N$ games the emergence of partial pooling equilibria becomes increasingly likely as N increases. Argiento et al. (2009) confirm that babbling equilibria emerge in games where states are not equiprobable, that synonyms arise and persist in games with too many signals and that categories typically appear in games with too few.

One thing the reinforcement learning framework does is to provide a robustness check to the evolutionary models described. Robustness of a result is checked by altering various aspects of a model and seeing whether the results still hold up.[22] The idea is usually to show that the causal dependencies identified in a model are not highly context-specific. In this case, we see that

---

[21] It is derived from psychological work on learning in animals by Herrnstein (1970), who in turn was inspired by the work of Thorndike (1898). It is sometimes called *Roth–Erev reinforcement learning, Herrnstein learning,* or *vanilla reinforcement learning.*

[22] The robustness of results, here, is not entirely surprising. The replicator dynamics are the mean field dynamics of basic reinforcement learning, meaning they track the expected average change of the learning dynamics (Hopkins, 2002).

the emergence of signaling is robust to changes in the population size and changes in the underlying evolutionary process. Furthermore, a host of variations on reinforcement learning yield similar outcomes (Barrett, 2006; Barrett and Zollman, 2009; Skyrms, 2010b; Huttegger and Zollman, 2011b), as do investigations of other sorts of learning rules (Skyrms, 2010b, 2012; Huttegger et al., 2014) and other types of signaling scenarios (Alexander, 2014).[23]

In addition to showing robustness, reinforcement learning models lend themselves to explorations of new possibilities that would be difficult to tackle in infinite population models. As noted, Skyrms (2010b) develops a simple model where actors invent new signals.[24] One interesting result of this model is that the invention feature seems to prevent the emergence of suboptimal signaling. Instead, actors who start with no meaning and no signals develop highly efficient patterns of behavior. Barrett and Zollman (2009) use the reinforcement framework to incorporate forgetting into evolution and to show that forgetting can actually improve the success of signal learning. So, in sum, we see that this learning framework extends our understanding of signaling both by opening up new possibilities for exploration and undergirding previous discoveries.[25,26]

3.3 Applying the Model

We have now seen that a great deal of theoretical progress has been made understanding the Lewis signaling game and the evolution of simple, common interest signaling. Let's turn to how philosophers of biology have applied the model to deepen our understanding of specific phenomena in the biological world. We will not be able to discuss all the literature that might be relevant. Rather, the goal is to see some examples of what the common interest signaling game can do.

---

[23] Blume et al. (1998) and Bruner et al. (2018) even find that in experiments of small groups playing the Lewis signaling game, actors tend to learn to signal, though this is less likely with more states and unequal probabilities of states. This result shows a robustness of signaling results in a very different sort of environment altogether.

[24] See also Alexander et al. (2012).

[25] Jeffrey Barrett has also used the reinforcement learning paradigm to go far beyond questions related to the emergence of basic meaning. His program explores questions related to epistemology, logic, and the evolution of cognition. He asks questions like, How might a notion of truth evolve (Barrett, 2016, 2017)? How might humans develop dependable epistemic networks (Barrett and Skyrms, 2017)? How can basic logical rules emerge (Barrett, 2013a, 2014c,a)? How can simple, but successful, epistemic language emerge (Barrett, 2013b, 2014b)?

[26] Another type of agent-based signaling model involves network analysis. In these models, agents typically learn from and interact with only those close to them in the network. Zollman (2005) and Mühlenbernd and Franke (2012) show how such models can develop regional meaning – in some areas, actors use one signaling system, and in other areas, the other. Wagner (2009) shows that this regional meaning tends to be optimal.

3.3.1 Animal Communication

As Skyrms (1996) points out, birds do it, bees do it, even monkeys in the trees do it – we have noted that signaling is rampant in the animal kingdom. Evolutionary results from signaling games help explain this ubiquity, but can they provide insights into particular cases?

Skyrms considers vervet alarm calls. In this case, signaling is costly – an alarm draws attention right at the moment a dangerous predator is nearby. This means that we should not expect signaling to emerge in the basic Lewis game. But the framework also elucidates what is missing – in many cases, vervets are signaling to their own kin, meaning that altruistic signalers are also creating benefits for other altruistic signalers.[27] As he shows, if there is even a small amount of correlation between strategies so that the benefits of signaling fall on signalers, communication can emerge.

Donaldson et al. (2007) ask why many animal alarm calls refer to a level of urgency perceived by the speaker, rather than to clear external world referents (such as hawks or snakes). They consider signaling games with fewer signals than states, where there are payoff reasons why some states might be categorized together. As they show, directing successful action, rather than referring to distinct states, is the primary driver of categorization in these games. Thus alarm calls are doing the fitness-enhancing thing by directly referring to the action that needs to be performed.

Barrett (2013a) and Barrett and Skyrms (2017) consider experimental evidence showing that scrub jays and pinyon jays infer transitive relationships based on learned stimuli. (For example, if they learn that blue > red and red > yellow, they are able to infer that blue > yellow.) They use reinforcement learning simulations to consider (1) how organisms might evolve a tendency to linear order stimuli in the first place and (2) how they might apply this tendency to novel scenarios. The signaling model illuminates how this sort of tendency might emerge and how it might later help organize information from the world. In other words, it gives a how-possibly story about the emergence of transitive inference in animals like jays. (More in a minute on how signaling models can be used to represent cognition.)

3.3.2 Human Language

Signaling games provide a useful framework for exploring human language. There are many features of language that go far beyond the behavior that

---

[27] Altruism, here, should be understood in the biological sense. These signalers decrease their fitness to improve the fitness of the receivers.

emerges in highly simplified game theoretic models, but nonetheless, progress can be made by starting with these simple representations.[28]

Payoff Structure, Vagueness, and Ambiguity

Jäger (2007) extends an idea we saw in Donaldson et al. (2007) – that, in many cases, there are payoff reasons why actors categorize together states of the world for the purposes of signaling. He develops a framework, the *sim-max game*, to explore this idea in greater detail.[29] Sim-max games are common interest signaling games where (1) actors encounter many states that bear similarity relations to each other, (2) actors have fewer signals than states, and, as a result, (3) actors categorize groups of states together for the purposes of communication. Similarity, in these models, is captured by payoffs. States are typically arrayed in a space – a line or a plane, say – where closer states are more similar in that the same actions are successful for them. Remember the mother and daughter signaling about the weather? In a sim-max game, the states might range from bright sun to a hurricane. Umbrellas would be successful for multiple states in the rainy range and a picnic for multiple states in the sunny range.

Several authors have used the model to discuss the emergence of vague terms – those with borderline cases, such as "bald" or "red." Vague terms provide an evolutionary puzzle. They are ubiquitous, but they are never efficient in the sense that linguistic categories with clear borderlines will always be better at transferring information (Lipman, 2009). Franke et al. (2010) use a sim-max model to show how boundedly rational agents – those who are not able to calculate perfect responses to their partners – might develop vague terms.[30] O'Connor (2014a) takes a different tack by showing how vagueness in the model, while not itself fitness enhancing, might be the necessary side effect of a learning strategy that is beneficial: generalizing learned lessons to similar, novel scenarios.

O'Connor (2015) also uses the model to address the evolution of ambiguity – the use of one term to apply to multiple states of the world. She proves that whenever it costs actors something to develop and remember signals, optimal languages in sim-max games will involve ambiguity, even though

---

[28] Zollman and Smead (2010) also use a signaling model to look at possible origins of human communication.

[29] It bears many similarities to the widely studied conflict of interest signaling game introduced in economics by Crawford and Sobel (1982).

[30] Similarly, Franke and Correia (2017) show that when actors imitate each other imprecisely, terms emerge that are vague across the group.

communication would always be improved by adding more signals.[31] In both of these cases, evolutionary models of signaling help shed light on linguistic traits that are otherwise puzzling.

Compositionality

One of the most important features of human language is compositionality – the fact that words can be composed to create new, novel meanings. For example, we all can understand the phrase "neon pink dog," even if we have never encountered this particular string of words.

Barrett (2007) provides a first model, where two agents send signals to a receiver who chooses actions based on the compositions of these signals.[32] As he shows, meaning and successful action evolve easily in this model. However, Franke (2014, 2016) points out that actors in Barrett's model do not creatively combine meanings but rather independently develop actions for each combination of signals. He presents a model with a sim-max character, where actors generalize their learning to similar states and signals. As he demonstrates, extremely simple learners in this model can creatively use composed signals to guide action. Steinert-Threlkeld (2016) also provides support for this claim by showing how simple reinforcement learners can learn to compose a negation symbol with other basic symbols. In doing so, they outperform agents who learn only non-compositional signaling strategies. Barrett et al. (2018) develop a more complex model where two senders communicate about two cross cutting aspects of real-world states, but an "executive sender" also determines whether one or both of the possible signals are communicated. The receiver may then respond with an action appropriate for any of the combined states or just one aspect of the world. In all, the signaling game here provides a framework that helps researchers explore (1) what true compositionality entails, (2) what sorts of benefits it might provide to signalers, and (3) how it might start to emerge in human language.[33]

3.3.3 Signaling in the Body

Within each of our bodies neurons fire, hormones travel through the blood stream, DNA carries genetic information, and our perceptual systems mediate

---

[31] Santana (2014) addresses the evolution of ambiguity using a different signaling model: one where receivers are able to use context to limit the possible meanings of a message they receive.

[32] See also Barrett (2009) and Nowak and Krakauer (1999). Komarova et al. (2001) use signaling models to discuss universal grammar.

[33] One last place where the signaling game has proved useful to the study of human language is to linguistic *pragmatics*. For space reasons, I do not discuss this literature at length. For more, see Van Rooij and Franke (2015), Franke et al. (2009), Jäger (2012), and Franke (2013b).

action with respect to inputs from the outside world. Each of these is an example of a process where information of some sort is transferred. Each might be represented and understood via the signaling game. There is an issue, though. When it comes to signaling within the body, certain features of the signaling model may not be present. Can we still apply the Lewisean framework?

Godfrey-Smith (2014) points out that there are many systems throughout the biological world that have partial matches to the signaling game model. He discusses, in particular, the match with two sorts of memory systems – neuronal memories and genetic memory. In the case of neuronal memory, there are clear states experienced by an organism, and a clear moment in which these states are written into synaptic patterns as a signal. However, there is arguably no clear receiver since synaptic weights directly influence future behavior.[34] Instead, we should think of these as write-activate systems, which have a partial match to the Lewis model. In the case of genetic signaling, various authors have argued that genes present a clear case of information transfer and signaling between generations (Shea, 2007, 2013; Bergstrom and Rosvall, 2011). Godfrey-Smith argues that this is also only a partial match, since there is no clear sender of a genetic signal. He dubs this an evolve-read system. More generally, we can think about many cases as having these sorts of partial matches to the full Lewis model.

In this vein, O'Connor (2014b) uses the sim-max game to model the evolution of perceptual categories. Every animal mediates behavior via perception. In this way, perceptual systems are signals from an animal to itself about the state of the world around them. In O'Connor's models, a state is a state of the external world, the action is whatever an organism does in response to that state, and the signal consists in mediating perceptual states within the body. As noted, sim-max games tend to evolve so that whatever states may be responded to similarly are categorized together. O'Connor argues that for this reason, we should expect categories to evolve to enhance fitness, not necessarily to track external similarity of stimuli.

Calcott (2014) shows how signaling games can be used to represent gene-regulatory networks: systems that start with environmental inputs (such as the presence of a chemical) and use chemical signals to turn genes on and off that help the organism cope with the detected state. Variations of the model involving multiple senders, and a receiver who can integrate this information can

---

[34] Relatedly, Cao (2012) looks at the possibility of using signaling games to represent information transfer between individual neurons in the brain. She argues that this is not a useful way to think about things because neurons do not have individual interests and do not have the capacity to behave flexibly in response to some signal they might receive. See also Godfrey-Smith (2013).

help explain how gene regulatory networks perform simple computations. Furthermore, the signaling structure induces a highly evolvable system – a new environment might be mapped to a signal to induce a preevolved action appropriate for it, for example. In a related project, Planer (2014) uses signaling games to argue that a typical metaphor in biology – that genes execute something like a computer code – is misleading. He draws out disanalogies between genetic expression and the execution of an algorithm. He then argues that genes are better thought of as the senders and receivers of transcription factors.

What we see from these examples is that even when there is not an entirely explicit signaling structure, the signaling game framework as developed by philosophers of biology can be extremely useful. It can help answer questions such as, Do our perceptual categories track the world? How can gene regulatory networks flexibly evolve? And, are genes really like computer programs?

As we have seen in this section, the common interest signaling game first entered philosophy through attempts to answer skeptical worries about linguistic convention. But the model has become a central tool in philosophy of biology, used to elucidate a range of topics. In addition, philosophers have helped develop an extended, robust understanding of the dynamics of signaling evolution. In the next section we will see what happens when we drop a key assumption from the Lewis model: that the interests of signalers always coincide.

4 Conflict of Interest Signaling

The Lewis games of the last section modeled signaling interactions where agents shared interests. Sender and receiver, in every case, agreed about what action should be performed in what state of the world. But what if sender and receiver are not on the same page? Remember the peacock mantis shrimps from the introduction, who would both like to reside in the same cozy burrow, and who use the colors on their meral appendages as signals to help determine who will get to do so? These actors have a clear conflict of interest – each prefers a state where he is the resident of the burrow, and the other is not. There might be some temptation to think, then, that successful communication should not be possible between these shrimp. Why should they transfer information to an opponent? And alternatively, why would they trust an opponent to transfer dependable information when their interests do not line up? Clearly, though, the mantis shrimp do manage to signal, as do many other organisms who have some conflict of interest – predators and prey, potential mates, etc.

We will begin this section with some of the most important insights in the study of conflict of interest signaling. In particular, both biologists and economists have used signaling models to show how signal *costs* can maintain information transfer when there are partial conflicts of interest. This insight is sometimes known as *costly signaling hypothesis*. There are issues with this hypothesis, though. First, empirical studies have not always found the right sorts of costs in real-world species. Second, on the theoretical side, because such costs directly detriment fitness, they seem to pose a problem for evolution.

Philosophers of biology have contributed to these theoretical critiques. But they have also explored ways to circumvent these worries and revamp costly signaling hypothesis. This section will discuss several ways in which philosophers have done this (1) by looking at *hybrid equilibria* – ones with low signal costs and partial information transfer, (2) by showing how communication that is even partially indexical (or inherently trustworthy) can lower signal costs, and (3) by showing that some previous modeling work has overestimated the costs necessary for signaling with kin.

At the end of the section, we will turn to a philosophical consideration in the background of this entire section: what is conflict of interest in a signaling scenario, and how might we measure it? As will become clear, attempts to answer this question press on a previous maxim: that when there is complete conflict of interest, signaling is impossible.

4.1 Costs and Conflict of Interest

The first thing to observe is that even the mantis shrimp do not have completely opposing interests. In particular, neither of them wants to fight, meaning that instead, they have a *partial conflict of interest*. The most classic case of partial conflict of interest signaling in biology comes from mate selection. In peacocks, for example, the peahen shares interests with high quality males because they would both like to mate. But while low-quality males would like to mate with her, she would not choose them. High-quality males could send a signal of their quality to facilitate mating, but low-quality males would be expected to immediately co-opt it. Analogous situations crop up in human society. The classic case is one of a business trying to hire. They share interests with high-quality candidates but have a conflict with low-quality ones. How can the business trust signals of quality it receives, given that low-quality candidates want to be hired?

Both biologists and economists figured out a solution to the signaling problem just introduced. In biology, Amotz Zahavi introduced what is called the *handicap principle*, which was later modeled using signaling games by Alan Grafen (Zahavi, 1975; Grafen, 1990). The insight is that communication can remain honest in partial conflict of interest cases if signals are costly in such a way that high-quality individuals are able to bear the cost, while low-quality

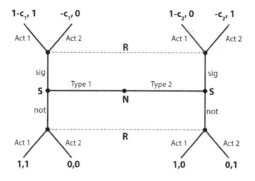

Figure 4.1 The extensive form respresentation of a partial conflict of interest signaling game.

individuals are not. In an earlier model, economist Michael Spence showed the same thing for the job market case described above (Spence, 1978). If a college degree is too costly for low-quality candidates to earn, it can be a trustworthy signal of candidate quality. To be precise, in these models, signaling equilibria exist where high-quality candidates/mates signal, and low-quality ones do not because of these costs.

For the purposes of this Element, I will describe a very simple version of a partial conflict of interest signaling game (borrowed from Zollman et al. (2012)). Figure 4.1 shows the extensive form. Nature plays first and selects either a high-type (type 1) or low-type (type 2) sender. Then the sender decides to signal or not. Upon observing whether a signal was sent, the receiver chooses action 1 (mate, or hire) or action 2 (do not mate, do not hire). The sender's pay-offs are listed first – they get a 1 whenever the receiver chooses action 1, minus a cost, c, if they signaled. We assume that $c_1 < c_2$, or high-type senders pay less to signal. Receivers get 1 for matching action 1 to a high-type sender or action 2 to a low-type sender. This corresponds, for example, to a peahen mating with a healthy peacock or rejecting an unhealthy one.

Consider the strategy pairing where high types send, low types do not, and receivers only take action 1 in response to a signal. This is not an equilibrium when costs are zero. This is because if a receiver is taking action 1 in response to the signal, low types will switch to send the signal. Once both high and low types send, there is no longer any information in the signal, and receivers should always take whatever action is best for the more common type.

Suppose instead that $c_1 < 1 < c_2$. High types gain a benefit of $1 - c_1$ from signaling if receivers are then willing to choose action 1 as a result. Low types cannot benefit from signaling, since the cost of the signal is more than the benefit from being chosen as a mate/colleague. Thus there is a *costly signaling equilibrium* where the signal carries perfect information about the type of the sender, and receivers and high types get a payoff as a result. Grafen (1990) showed that this type of equilibria are ESSs of costly signaling models.

It has been argued that costly signaling equilibria can help explain a variety of cases of real-world signaling. The original application for biology, as noted, is to the sexual selection of costly ornamentation. The peacock's tail, for example, imposes a cost that only healthy, high-quality peacocks can bear. Contests, such as those between mantis shrimp, are another potential application (Grafen, 1990). If it is costly to develop expensively bright meral spots, this may be a reliable indicator of strength and fitness. Another potential application regards signaling between predator and prey. Behaviors like stotting, where, upon noticing a cheetah, gazelles spring into the air to display their strength, may be costly signals of quality.[35] And as we will see in Section 4.4, similar models have been applied to chick begging.

There are a few issues, though. In many species, when the actual costs paid by supposedly costly signalers are measured, they come up short. They do not seem to be enough to sustain honest signaling (Caro et al., 1995; Silk et al., 2000; Grose, 2011; Zollman et al., 2012; Zollman, 2013). This has led to widespread criticism of the handicap theory.

There are theoretical issues with the theory as well, which arise because costs are generally bad for evolution. Bergstrom and Lachmann (1997) point out that some costly signaling equilibria are worse, from a payoff standpoint, than not signaling at all. And once one does a full dynamical analysis, it becomes clear that the existence of costly signaling equilibria is insufficient to explain the presence of real-world partial conflict of interest signaling. This is because costly signaling equilibria tend to be unlikely to evolve due to the fitness costs that senders must pay to signal. These equilibria have small basins of attraction under the replicator dynamics (Wagner, 2009; Huttegger and Zollman, 2010; Zollman et al., 2012). Is the costly signaling hypothesis bankrupt? Or does it nonetheless help explain the various phenomena it has been linked to? Let us now turn to further theoretical work that might help circumvent these worries.

4.2 The Hybrid Equilibrium

By the early 1990s, economists had identified another type of equilibrium of costly signaling models (Fudenberg and Tirole, 1991; Gibbons, 1992). In a *hybrid equilibrium*, costs for signaling are relatively small. But even small costs, as it turns out, are enough to sustain signals that are partially informative.

---

[35] This is controversial, though (Caro, 1986; Fitzgibbon and Fanshawe, 1988). See Johnstone (1995) for an overview of empirical support for the costly signaling hypothesis.

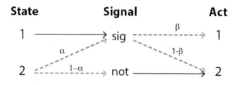

Figure 4.2 The hybrid signaling equilibrium.

Let's use the game in Figure 3.1 to fill in the details. Suppose the cost for low types, c_2, is constrained such that $0 < c_2 < 1$, and the cost for high types is less than this, $c_1 < c_2$. Under these conditions, there is a hybrid equilibrium where (1) high types always signal, (2) low types signal with some probability α, (3) the receiver always chooses action 2 (i.e., do not mate) if they do not receive the signal, and (4) the receiver chooses action 1 (i.e., mate) with probability β upon receipt of the signal. Figure 4.2 shows this pattern.

The values for α and β that sustain this equilibrium depend on the details of the costs and the prevalence of high types in the population. If x is the proportion of high types, $\alpha = x/(1-x)$ and $\beta = c_2$ (Zollman et al., 2012). These values are such that low types expect the same payoffs for sending and not sending (and thus are willing to do both with positive probability), and high types expect the same payoffs from taking either action given the signal. Unlike the costly signaling equilibrium, only partial information is transferred at the hybrid equilibrium. Upon receipt of the signal, the receiver cannot be totally sure which state obtains. But she is more certain of the correct state than she was before observing the signal and can shape her behavior accordingly.

In some ways, the hybrid equilibrium seems inferior to the costly signaling one – in the latter perfect information is transferred. In the former, high types and receivers are only somewhat able to act on their shared interests. From an evolutionary perspective, though, the hybrid equilibrium proves very significant. Unlike the costly signaling equilibria, low signaling costs mean that hybrid equilibria are often easier to evolve in costly signaling models (Wagner, 2013b; Huttegger and Zollman, 2010; Zollman et al., 2012; Kane and Zollman, 2015; Huttegger and Zollman, 2016). This might help explain how partial conflict of interest communication is sustained throughout the biological world. Small costs help maintain information transfer, without creating such an evolutionary burden that communication is not worthwhile. Furthermore, these low costs seem more in line with empirical findings in many cases.[36]

---

[36] Rubin et al. (2015) show that when signal costs are in the right range for the hybrid equilibrium, groups of humans indeed develop partially honest communication patterns.

4.3 When Honesty Met Costs: The Pygmalion Game

Hybrid equilibria can evolve and guarantee partial information transfer. There is another solution, though, to the problem posed by the costs of signaling. Philosophers of biology have argued that, in fact, the very setup of the costly signaling game builds into it an unrealistic assumption.

Aside from signal cost, there is another way to guarantee information transfer in partial conflict of interest scenarios. This is for high types to have a feature or signal that is unavailable to low types, or unfakeable. Suppose humans were able to innately read the quality of job market candidates, for instance. There would be no need for expensive college degrees to act as costly signals, because it would be obvious which candidates to hire. A feature that correlates with type in this way is sometimes called an *index* rather than a signal (Searcy and Nowicki, 2005).[37]

Huttegger et al. (2015) point out that while biologists have traditionally thought of costly signaling and indexical signaling as alternative explanations for communication in partial conflict of interest cases, this distinction is a spurious one. They introduce a costly signaling model called the *pygmalion game* where, when low types attempt to signal, they do not always succeed. With respect to stotting, this would correspond to fast gazelles having a higher probability of jumping high enough to impress a cheetah than slow ones. In other words, the signal is partially indexical. As they point out, at the extremes, their model reduces to either a traditional costly signaling model – if low and high types are both always able to signal perfectly – or to an index model where only high types are able to send the signal at all.

Between these extremes is an entire regime where costs and honesty trade-off. As they show, the harder a signal is to fake, the lower the costs necessary to sustain a costly signaling equilibrium. The easier the signal is to fake, the higher the costs. In these intermediate cases where signal costs are not too high, the signaling equilibria often have large basins of attraction. So if signals are even a bit hard to fake, the costs necessary to sustain full, communicative signaling might not be very high, easing the empirical and theoretical worries described earlier in the section.[38]

---

[37] In fact, although I mentioned stotting earlier as an example of a costly signal, many think that this is actually an index (Fitzgibbon and Fanshawe, 1988).

[38] Zollman (2013) gives a nice overview of criticisms of the handicap principle and further alternative explanations for partial conflict of interest signaling. See also Bergstrom and Lachmann (1998).

4.4 Altruism, Kin Selection, and the Sir Philip Sydney Game

As mentioned earlier, the costly signaling hypothesis may also explain elements of chick begging. Maynard-Smith (1991) introduced the *Sir Philip Sydney game* to capture this case.[39] In this model, chicks are needy or healthy with some probability. They may pay a cost to signal (cheep), or else stay silent. A donor, assumed to be a parent, may decide whether to pay a cost to give food conditional on receipt of the signal. If chicks do not receive a donation, their payoff (chance of survival) decreases by some factor that is larger for needy chicks. Notice that this model is analogous to the partial conflict of interest models described thus far, but with a notable difference – healthy and needy chicks do not pay differential costs to signal but instead receive differential benefits.

In this model, there is never a direct incentive for the receiver to donate food, since this only leads to a direct cost. Any such donor would have to be an altruist in the biological sense, i.e., willing to decrease their fitness in order to increase the fitness of a group member. But, as Maynard-Smith points out, in cases where donor and signaler are related, donation may be selected via kin selection. He uses a heuristic for calculating payoff in this model, which supposes that the donor (receiver) gets a payoff themselves when they donate to a needy chick related to them (more on this shortly). As Maynard-Smith shows, if there is total relatedness between donor and recipient, the signaling equilibrium is an ESS – only needy individuals send the signal, and the receiver donates in response. If there is partial relatedness, signal costs can ensure a signaling equilibrium as well. Only needy individuals are willing to pay the costs to signal, and the receiver is only willing to donate upon receipt of the signal. As Johnstone and Grafen (1992) point out, the higher the relatedness, the lower the cost necessary to protect signaling. As such, relatedness provides another solution to worries about costly signaling by increasing shared interests between signalers.

Furthermore, Bruner and Rubin (2018) give a methodological critique of the heuristic used by Maynard-Smith (1991) to analyze the Sir Philip Sydney game that further addresses these worries. Maynard-Smith's heuristic calculates *inclusive fitness*, which takes into account how an individual influences the fitness of group members who share genetics. In particular, Maynard-Smith calculates payoffs to an individual and adds to this the payoffs of relatives weighted by relatedness. Although it is widely recognized that this is not the correct way to calculate inclusive fitness, the heuristic is used because it is simple and captures the basic idea that inclusive fitness should take influence

---

[39] Why the name? As Maynard-Smith writes, "It is reported that Sir Philip Sydney, lying wounded on the battlefield at Zutphen, handed his water bottle to a dying soldier with the words 'Thy necessity is yet greater than mine'" (1034).

on relatives into account.[40] What Bruner and Rubin show is that under more proper calculations, honest signaling emerges in the Sir Philip Sydney game even without costs, for partially related individuals. In other words, their results more strongly support the claim that relatedness can sometimes solve the costly signaling problem.[41]

4.5 Measuring Conflict of Interest

While this entire section has focused on partial conflict of interest signaling, we have not given a clear definition of what conflict of interest is. Complete common interest means that sender and receiver perfectly agree about which outcomes of the game are preferable to which others. In conflict of interest scenarios, this is no longer true. But while it is clear that interests can conflict to greater or lesser degrees, it is not entirely clear how this should be measured. One typical claim in signaling theory is that signaling cannot occur without some common interest and that the more shared interests there are, the easier it is to generate signaling. Without a measure of conflict, though, these claims are hard to assess.

Godfrey-Smith and Martínez (2013) wade in. They consider a suite of randomly generated 3×3×3 signaling games, and for each one, calculate the degree of common interest between sender and receiver. They compare preference orderings over outcomes and generate a measure between 0 and 1 that ranks how different these are. They also calculate, for each game, whether at least one equilibrium exists where information is transfered and where there are no signal costs whatsoever. Surprisingly, they find that for some games with zero common interest information transfer is possible at equilibrium, though this is more likely the greater the common interests are.[42] Martínez and Godfrey-Smith (2016) use replicator dynamics models to derive similar, but explicitly evolutionary, results – that common interest strongly predicts whether a population playing a game is likely to evolve communication, but sometimes this happens despite zero common interest.

This finding is dependent on their measure, though. If they took zero common interest to correspond to a zero-sum game (where more for you is always less for me), the finding would not hold. Surprisingly, though, Wagner (2011) does show how in a zero-sum model, information can be transfered in evolving signaling populations. This is not equilibrium behavior but occurs as a result

---

[40] Grafen (1982) shows why it is not actually correct. See Skyrms (2002a) and Okasha and Martens (2016) for more.

[41] See Rubin (2018) for further critiques of inclusive fitness and debates surrounding it.

[42] Conversely, some games with high common interest on their measure have no such equilibria.

of chaotic dynamics.[43] What we see here is that theoretical work on the meaning of conflict may have important consequences for signaling theory. There is certainly room for further elaboration of this concept.

The costly signaling hypothesis has been beset by critics, but philosophers of biology have helped contribute to the literature showing why costs may nonetheless help sustain partial conflict of interest signaling. Even low costs can work for partial information transfer, especially when signals are partially indexical or when organisms are related. Furthermore as we saw, philosophers of biology are contributing to conceptual work defining what conflict of interest signaling is, and, relatedly, assessing claims about the relationships between conflicts, costs, and information transfer.

The next section continues with work on the signaling game but turns toward more philosophical issues related to signal meaning.

5 Meaning and Information in Signals

A vervet monkey sees an eagle and sends an alarm call. The rest of the troupe scatters. As we have seen, this sort of interaction can be modeled using a signaling game. And these models help us understand questions like, Why do the vervets act as they do? How might they have evolved this system of behaviors? What benefits do they derive from signaling?

There are further questions we might ask, though, of deep philosophical significance. What does the vervet signal mean? Does it have a meaning similar to the human term *eagle*? Or maybe *run*? Or should meaning in animal signaling be understood as totally distinct from human language? What about when the vervet signals and there is no eagle there? Is this lying? Deceit? What about, say, molecular signals between bacteria, or hormonal signals within the body? Do these have meanings? Could they be deceitful? In general, what is the content of different signals ranging over the animal kingdom, and where does this content come from?

Some philosophers have used signaling games as a framework to try to answer these questions. In particular, models can help precisify what one might mean by content in various cases. The goal of this section will be to discuss issues related to meaning, content, information, and representation as they have been addressed using games. This discussion sits within an enormous philosophical literature about meaning, reference, and semantics. For space reasons, I ignore most of this literature and focus on work that employs signaling games to discuss meaning and information.

---

[43] Smead (2014a) also looks at models ranging from common to conflict of interest to consider what conditions might be favorable to organisms who adopt behavioral plasticity or learning.

The section starts with a brief introduction to information, in the information theoretic sense. We'll then turn to accounts of meaning in signals that draw on information theory and to accounts that otherwise ground their notion of meaning in games and equilibria. This will lead into a very brief treatment of teleosemantic theories of meaning. After that, we will look at some more specific issues: Do animal signals have indicative content, or imperative content, or both? How can we tell? Can non human signals include misinformation? Can they be deceptive? And what makes them so?

5.1 Information Theory

A good place to start, in thinking about theories of meaning and content, is with the theory of *information*. Information theory, beginning with the influential work of Claude Shannon, focuses on questions regarding how much information can be sent over various channels, and how best to package it (Shannon, 1948). This theory also gives a way to measure information working off of probabilistic correlations.

It might sound strange to talk about measuring information, but we can think of it in the following way. The more something reduces the uncertainty of an observer, the more information it contains. Consider a random coin flip. Before observing the coin, you are uncertain about which face will be up. There is a 50/50 probability of either heads or tails. Upon observing the coin, your uncertainty is reduced. You are now certain that it is, say, heads. Consider a different case, where you are flipping a biased coin that comes up heads 99 percent of the time. Now even before you flip, you're almost certain that it will come up heads. Once you observe that the coin is, indeed, heads, your uncertainty is reduced, but only by a bit. Alternatively, consider throwing an unweighted die. There are six sides, each of which will turn up with probability 16.6 percent. Now there is more uncertainty than in the unbiased coin case. With the coin you knew it would be either heads or tails, but now it could be any of six outcomes. When you observe that the face is "four," you reduce your uncertainty more than with the coin.

Shannon (1948) introduced the notion of *entropy* to capture the differences between these cases. Entropy tells us how much uncertainty is reduced, and thus how much information is conveyed through some *channel* on average. A channel is some variable that can be in multiple states – a flipped coin, a die, or else a monkey alarm, or a job resume stating whether a candidate attended college. If the probability of each possible state i is p_i, the entropy of a channel is

$$\sum_i -p_i\log_2(p_i). \tag{5.1}$$

Let's break this down. For each possible state of the channel, we take its probability. In the case of the coin, $p_H = .5$ and $p_T = .5$. Then we multiply this by the negative log of that probability and add it up. In the coin case, $-\log_2(.5) = 1$. So the entire formula here yields $.5 * 1 + .5 * 1 = 1$. When we observe the coin, we gain one bit of information.

Why the logarithm part? The answer is that it has nice properties for measuring information. The log of 1 is 0. Intuitively, if something has a probability of 1, i.e., is entirely certain to happen, our uncertainty is not reduced at all from observing it. On this measure, it carries no information. As the probability of an event decreases toward 0, the negative log becomes larger and larger. This, again, makes intuitive sense: if something is highly unlikely, we learn a lot from observing it. The more unlikely the event is, the more information it carries on this measure. We could use another logarithm, if we wanted, which would simply change the unit of information from bits to some other measure. (And, to be clear, the *negative* log is used simply so that the measure will be positive.)[44]

Let's try this for the biased coin. The probabilities are $p_H = .99$ and $p_T = .01$. If we take the negative log of these, we get $-\log_2(.99) = .014$ and $-\log_2(.01) = 6.64$. Notice that we get very little information from heads and quite a lot from tails. The entire formula is thus $.99 * .014 + .01 * 6.64 = .08$. This is much less than the expected information in the equiprobable coin, tracking our intuition that we don't learn much in this case (even though we learn a lot in the unlikely case that tails comes up). For the die roll the probability is $.166$, and the negative log of that is 2.6, so the entropy is 2.6, tracking the intuition that we expect to learn more.

How does this all connect up with signaling and communication? We can use information theoretic measures to tell us not just how much information is transmitted on average via some channel but how much that channel tells us about some other distribution. If our channel is vervet alarm calls, for example, we can measure how much these alarms tell us about the actual state of the world vis-à-vis predators.

As Skyrms (2010a, 2010b) shows, the signaling game framework provides a nice way to do this, using what is called the *Kullback–Liebler divergence*.[45] This measure can tell us for any two distributions of events, such as states of predators in the forest, and alarm calls given, how different the second is

---

[44] One more note. Although $-\log_2(0)$ is undefined, we stipulate that a signal with no probability of being sent has 0 information value.

[45] Some others have depended on the notion of *mutual information* between two variables. This measure tracks the amount of information one gains about a variable (the state) by observing the other variable (the signal).

from the first. Let's transition to the language of signaling games to understand this measure better. Let p_i be the prior probability of the state and p_i^{sig} be the probability of that state after the signal arrives. The measure is

$$\sum_i p_i^{\text{sig}} \log_2(p_i^{\text{sig}}/p_i). \tag{5.2}$$

Here is one way to think about this formula. In trying to figure out how much information is in the signal, we want to know how much it changes the probability of the state. So we can take the ratio of the new and old probabilities and take the log of that, $\log_2(p_i^{\text{sig}}/p_i)$. This gives us the information in the signal about state i. Then we can average this over all the states to get the full information in the signal about all possible states.

Let's apply this. Suppose I am flipping an unbiased coin. Also suppose that after I observe it, I turn on either a red light for heads or a blue light for tails. We know the initial probabilities are $p_H = .5$ and $p_T = .5$. Let us consider the information in the red light. If I am a perfect observer, when the light is red the new probabilities are $p_H^{\text{sig}} = 1$ and $p_T^{\text{sig}} = 0$. Then we can fill in the following values: $1*\log_2(1/.5)+0*\log_2(0/.5)$. This is equal to $1*1+0*-\infty$, or to 1. If the signal carries perfect information about the state, the amount of information in it is equal to the entropy of the state.

Suppose that I am an imperfect observer and that when the state is heads, I flip the red light switch 90 percent of the time, and the blue light 10 percent, and vice versa when the state is tails. Now when we observe the red light, we cannot be certain that the state was heads. Instead, the probability the state is heads is $p_H^{\text{sig}} = .9$ and for tails $p_T^{\text{sig}} = .1$ So now we get $.9*\log_2(.9/.5)+.1*\log_2(.1/.5)$ or $.9 * .85 + .1 * (-2.32) = .532$. There is less information about the state when the signal does not correlate perfectly with it.

We can use this kind of measure to calculate what we might call *natural information*. For instance, there is an old adage that "where there is smoke there is fire." If we like, we can figure out what the probability is that there is fire at any point, p_F, or no fire, $p_{\neg F}$. We can also figure out, given that there is smoke, what the probabilities are that there is or is not fire. From this, we can calculate how much information the smoke carries about fire.

We can also use this measure on acquired meaning. Skyrms (2010a, 2010b), in particular, uses this measure within the signaling game context. In a signaling model, the probabilities of the states, and of signals being sent given any state, are perfectly well defined. This means that we can measure the information in the signals about the states at any point. As Skyrms points out, we can watch signals go from carrying no information to carrying nearly full information about a state as a simulation progresses.

5.2 Semantic Content

Many have expressed dissatisfaction with the idea that we can say everything we'd like to about the meaning of signals using just information theory. Consider vervet signals. We can point out that there is a probabilistic correlation between the presence of a leopard and the vervet leopard alarm call. In this sense there is a measurable quantity of information in the signal about leopards. But we might also want to say that the signal has *semantic* content, i.e., that its content is inherently about something, much in the way that human language is. In this case, it seems natural to say that the content in the signal is very similar to the content in the human sentence "There is a leopard!" (In fact, in their landmark articles about vervet signaling, biologists Dorothy Cheney and Robert Seyfarth describe the vervets as engaged in "semantic communication" [Seyfarth et al., 1980a, 1980b].)

In the case of human language, one can appeal to intentions to specify semantic content (Grice, 1957, 1991). For example, when I say "duck!" it might either be that I intend the listener to know there is a duck there or am telling him to duck. My intentions define the content. In the case of vervets (or bacteria) it is not clear that they have intentions of this sort. But we still would like to say their signals have semantic meaning in many cases. The worry is that information theory gives us no good way to describe this sort of content. In other words, it can tell us how much, but not what about.

Contra these worries, Skyrms (2010a, 2010b) uses the signaling game framework, and information theory, to define semantic content in signals. In many ways, this account bears similarities to that of Dretske (1981), who uses information theory to develop an account of mental content and representation. On Skyrms's view, the content of a signal is a vector specifying the information the signal gives about each state. For states i, the vector is

$$< \log_2(p_1^{\text{sig}}/p_1), \log_2(p_2^{\text{sig}}/p_2), ..., \log_2(p_n^{\text{sig}}/p_n) > . \tag{5.3}$$

This is just the Kullback–Liebler divergence measured for each state the signal might tell us something about. To give an example, consider me flipping coins and turning on light switches again. Suppose there is an unbiased coin and perfect observation. The red light has the content $< 1, -\infty >$. Consider a similar case with a six-sided die and six perfectly correlated lights. The content in the yellow light (only and always sent in state 4) would be $< -\infty, -\infty, -\infty, 2.6, -\infty, -\infty >$. As Skyrms notes, this account is more general than traditional accounts of propositional content. On his account, we can sometimes think of these vectors as carrying propositional content, i.e., having the content of a proposition like "the coin flip was heads." The negative infinity entries tell us which states do not obtain, and the proposition is the conjunction of the remaining entries, i.e., "the state is 1, 4 or 6." When no states are ruled out, the signal has semantic informational content, just not propositional content.[46]

Godfrey-Smith (2011) suggests an amendment to this account. As he points out, we usually think of the content in a signal as about the world, not about how much the probability of a state was moved by a signal. His suggestion is that the proper content of a signal should be the final probabilities. So in the case of the die, the content would be $< 0, 0, 0, 1, 0, 0 >$, telling us that after observing the yellow light, it is perfectly certain the roll was four. This has a nice property. Say that we know ahead of time that the weather will be sunny and receive a signal that specifies the weather is sunny. On Skyrms' account, this message has no content because it does not change probabilities, but on Godfrey-Smith's account, it does.

Birch (2014) raises a problem for content in both views. As he points out, a key feature of semantic content is that it can be misleading or false. We can say "there is a leopard!" when there is no leopard, but nonetheless the statement still means that there is a leopard. But on these accounts, if a signal is sometimes sent in the "wrong" state, it also means that state, because it increases the probability of that state.

To see the problem more clearly, let's focus on propositional content. Signals will only be propositional on the Skyrmsian accounts when they are never sent in the wrong state. Consider a capuchin monkey who gives an alarm call in order to sneak food while dominant group members are hiding. It seems natural to say that the propositional meaning of the call is still "there is a leopard" and that the call is simply misleading. On the Skyrmsian view, though, the signal cannot have propositional content because it does not exclude a state. Instead, it would actually mean, to some degree, that there is not a leopard as well as that there is.

This is sometimes called the *problem of error*, and similar objections have been made to information theoretic accounts of mental content.[47] Birch points out that there are a few ways forward. One is to appeal to notions of function to ground content – the teleosemantic approach. Then, if a signal is used in a context that did not help select for it, or stabilize it in an evolutionary process, we can say it is misleading. (More on teleosemantics shortly.) The other is to specify fidelity conditions for a signal that define its content and allow one to

---

[46] In this vein, Ventura (2017) suggests that we stop using propositional content to account for rational behavior and focus on more general notions of content.

[47] Godfrey-Smith (1989) and Crane (2015) worry that the problem of error cannot be solved.

say when it is being used in a misleading way. The trick is to do this in a non arbitrary way.

Birch argues that evolutionary models of Lewis signaling games give a very natural way to do this. A population may be in a state that is close to a separating equilibrium. If so, on his account, the meaning of a signal is specified by its propositional content at the nearest equilibrium. So if the population is close to a state where an alarm is always sent when there is a leopard, the meaning of the alarm is leopard. When some individuals in the population send the alarm and no leopard is present, the fidelity criteria specified by the equilibrium do not hold, and thus the signal can be false.

There are various issues with defining closeness of equilibria. Skyrms and Barrett (2018) make another suggestion, which is that fidelity criteria might instead be defined by a fully common-interest interaction or "sub game" that stabilizes signaling. For example, capuchin monkeys will not usually be in a context where false alarms allow them to steal food. Under normal circumstances, they share interests that stabilize alarm signaling. These interactions define fidelity criteria. On the other hand, the contexts where interests diverge such that the subordinate capuchin is tempted to send a false signal do not define fidelity criteria.[48]

Shea et al. (2018) move in a similar direction. They do not hunt for a fully common-interest interaction on which to ground signal meaning. Instead, they separate informational content, which can be defined in a Skyrmsian manner, from functional content. Functional content, for them, exists in cases where a signal is improving the payoff of sender and receiver over a non signaling baseline and corresponds to the states where the signal is improving payoff in this way. In other words, the functional contents track the states that are part of selecting for a signal. Their functional content is also a vector (weighted to sum to 1) and so, like the other measures we have seen, ranges in degrees.

These suggestions share some features with a very influential branch of philosophical thought, to which I now, briefly, turn.

5.3 Teleosemantics

The teleosemantic program has a long, rich philosophical history. It is beyond the purview of this Element to survey this literature, but it is worth taking a little space to discuss.

---

[48] See also Franke (2013a).

The key idea behind teleosemantics is that signals derive their content from some sort of selective process.[49] In the case of evolution by natural selection, the content of a signal is intertwined with its function. If an alarm signal was selected to fulfill the function of warning group members in the presence of a leopard, its content comes from this process. The content helps explain why the signal has persisted through selection.

Millikan (1984), in her influential account, uses a framework that shares many properties with the Lewis signaling game. She refers to senders and receivers as "producers" and "consumers," and she argues that by dint of evolutionary processes whereby signals play a role in promoting successful behavior, these signals derive semantic content. For Millikan, much of this content has a particular character, which is that it both indicates some state of the world and suggests a responsive action (Millikan, 1995). In the leopard alarm case, we might say that the signal has a combined content of "there is a leopard" and "run up a tree." She calls this *pushmi-pullyu* meaning, after the llama-like creature of Dr. Doolittle with heads on both ends. From the perspective of the Lewis signaling game, this is an apt suggestion. In the basic models, signals at equilibrium are correlated both with state and action, so it is natural to think of them as possessing pushmi-pullyu content. As we will see in the next section, variations of the model can pull this content apart.[50]

Relatedly, (Harms (2004a, 2004b) has argued that there is no need to suppose that animal signals have propositional content. Instead, he proposes that signals of the type modeled in the Lewis game, with pushmi-pullyu meaning, have "primitive content." This simple content emerges as a convention develops for sender and receiver. As he points out, it is hard to translate this content to language – the phrase "there is a leopard" is full of extra meaning and implications, for example. Instead, primitive content, although it can be described with human language, is really content of a different sort, referring to state–act pairs.

5.4 Indicatives and Imperatives

As we have just seen, Millikan and Harms describe meaning with two kinds of content, one referencing the world and the other referencing the action that ought to be performed given the state of the world. These two types of content

---

[49] Much work in this literature focuses on mental representation rather than signaling per se. But, as noted, mental representations can be thought of as intra organism signals. For more, see Dretske (1981), Millikan (1984), and subsequent work.

[50] It is the teleosemantic framework that Cao (2012) uses to reject the sender-receiver framework as a good representation of neuronal signaling. Because neurons do not have interests or fitnesses, there is no receiver, and their signals cannot gain teleosemantic content in the right way.

are sometimes called *assertive* and *directive*, or else *indicative* and *imperative*. Lewis (1969), in fact, drew attention to this distinction as well, by making a divide between "signals-that" and "signals-to." As Millikan (1995) points out, some human phrases have pushmi-pullyu meaning. She gives a few examples, such as "The meeting is adjourned," which prompts people to leave while also making an indicative statement.[51] We have other phrases and terms where the content is mostly of one sort or the other. "Duck!" or "put your shoes on!" or "don't do that!" are all imperative-type statements, while "I am sleepy" or "that cat is fat" are indicative. Animal signals sometimes seem to share these distinctions. For California ground squirrels, one signal is usually given for aerial predators, which are far away and take time to approach, and another for ground predators, which are usually closer and require immediate action. However, the squirrels reverse the signals for aerial predators that are close and ground predators who are far. Thus the content of the signals does not nicely map to predator species. Rather, one is more like "a predator is present" and the other is more like "run!"

Huttegger (2007b) uses the signaling game to try to pull apart these two sorts of meaning. One way to drive a wedge is to consider who is deliberating (Lewis, 1969). If the sender deliberates over what to send, and the receiver does not deliberate over how to act, the signal can be thought of as imperative, and vice versa. Huttegger looks at a version of the Lewis game where in some states it is better for either the sender or receiver to deliberate and where their strategies allow them to choose whether to do so. As he shows, well-coordinated systems of behavior evolve with signals that have one type of content or the other.

Skyrms (2010a, 2010b) as noted, gives a measure of the information content in a signal about a state. I neglected to mention that he also gives a measure of the content about a act. This corresponds to how much the signal changes the probability that each possible act is performed. Zollman (2011) points out that Huttegger's model does not break informational asymmetry in the way we would want in pulling apart indicative and imperative meaning. Each type of signal carries full information about act and state at Huttegger's equilibrium. Zollman instead gives a model where a sender communicates with two receivers who must coordinate their acts. The sender can either send one message indicating the state, and the receivers figure out which action they must take, or else the sender can send two messages indicating the two ideal acts for

---

[51] Though Jonathan Birch, in personal correspondence, points out that in a case like this the imperative content is more pragmatic than semantic.

the receivers. This version of the model indeed breaks informational symmetry between indicative and imperative signals.[52]

In Section 3 we saw how Lewis games can be applied to understanding internal signaling. Martínez and Klein (2016) combine this insight with work on the indicative/imperative distinction to argue that pain is an internal signal with mostly imperative content. This is because pain, again, has an informational asymmetry. A similar pain can attach to many causes but directs just a few actions. It thus carries more information about acts than about states.

5.5 Deception

When a capuchin monkey sends an alarm call in order to snatch food from a dominant group member, is it "lying"? Is it deceiving? In the case of human language, we can again define deceit with respect to intention. The other day my daughter told me she had not drunk her sister's kombucha.[53] She had. And she intended me to not know that. Most animals do not have the cognitive capacity to intend to mislead in this way. Still, the capuchin case sounds like deception. Another classic case is that of firefly signaling. The genus *Photuris* has co-opted the mating signals of the genus *Photinus* to lead the latter species to their deaths. Unsuspecting fireflies head toward the signal, anticipating finding a willing mate, and are instead eaten. Isn't the *Photuris* signal deceptive?

Like the rest of this section, debates over deception in the animal kingdom are widespread and go well beyond the signaling game framework. Again, we focus on signaling games and on the philosophical literature. As we will see, though, even in this smaller literature, there is a great deal of debate. Readers may want to look at Maynard-Smith and Harper (2003) and Searcy and Nowicki (2005) for some wider perspective.

Before discussing deception, we will want to consider a related concept: misinformation. This is because many theorists define deception as a special case of misinformation. For Skyrms (2010b), misinformation occurs whenever a signal increases the probability of a state that does not obtain, or decreases the probability of the true state. Fallis and Lewis (2019) argue that this is too broad a notion, because sometimes misinformation in this sense increases the overall accuracy of beliefs (on a host of measures offered by formal epistemologists). Instead, we should use a good measure of accuracy (which one is still under hot debate) to define misinformation. Most theorists agree that this is not enough

---

[52] Franke (2012) also uses a signaling game to model cases of *dynamic meaning*, where imperatives actually change the state of the world as they are uttered because of social obligations. See also Blume and Board (2013).

[53] OK, whatever, we're from California.

for deception, though, and add conditions related to payoff, typically that the sender benefits, the receiver is hurt, or both.

Deception for Skyrms (2010b), for example, occurs when there is misinformation that also increases the payoff of the sender at the expense of the receiver.[54] Skyrms's account, though, fails to differentiate between two possible sorts of "deception" (Godfrey-Smith, 2011; Franke, 2013a; McWhirter, 2015; Fallis and Lewis, 2019; Martínez, 2019). In some cases, a sender can use ambiguity to manipulate receiver behavior. For instance, the sender might send *a* in both state 1 and state 2, but in no other states, in order to make the receiver choose an option the sender prefers. (For space reasons, I do not give a concrete example, but see Skyrms [2010b]; Godfrey-Smith [2011]; Franke [2013a]; McWhirter [2015], Fallis and Lewis [2019].) In such a situation, the probability of a non actual state is increased by the signal, and in a way that systematically advantages the sender and disadvantages the receiver. But the two are still involved in mutually beneficial equilibrium behavior, and furthermore, the signal can be thought of as half true in the sense that it correctly excludes states other than 1 and 2.[55] This looks quite different from the case where *Photuris* co-opts the cooperative signaling of *Photinus*. In the latter case, enough signaling by *Photuris* would make the signaling behavior collapse. This leads to a family of proposals from Godfrey-Smith (2011), Franke (2013b), Shea et al. (2018), and, ultimately, Skyrms and Barrett (2018). These proposals are approximately that the content of signals should be defined by their "sustaining" uses – the ones that keep the behavior going on an evolutionary scale – or else their common interest uses, or beneficial uses, and deception, in contrast, should be limited only to "unsustaining" uses, or ones that undermine the signal. For Shea et al. (2018), for example, deception occurs when a signal is sent with false functional content (in the sense outlined earlier) to the benefit of the sender.[56]

McWhirter (2015) worries about a different issue, which is that we often think about deception as requiring a deviation from expected behavior. For him,

---

[54] Lachmann and Bergstrom (2004) give an earlier proposal: deception occurs when *information value* is negative, i.e., it hurts payoffs for the reciever. This seems to define deception away, however, as they also prove that at any signaling equilibrium, the information value of the signals will be positive.

[55] As Fallis and Lewis (2019) point out, in many of these cases, there is no misinformation in their sense – receiver beliefs are more accurate after the signal. And Martínez (2019) makes clear that even the most "fair" outcomes between sender and receiver in some such games will be deceptive on Skyrms's definition, despite a high level of cooperation between sender and receiver at these outcomes.

[56] Martínez (2015) argues, contra these proposals, that there are intuitive cases of deception that are not non sustaining.

deception occurs only when an actor misuses a signal in a state, in the sense that they send it more often than the rest of the population, in a way that benefits themselves and hurts the receiver. On this definition, a signal cannot be universally deceptive but can only be deceptive with respect to some population level baseline. Fallis and Lewis (2019), contra this, point to some natural cases where behavior used across a population can look like deception. They also point out that an individual could send a signal that differed from the standard usage, to their benefit, but that improved the the epistemic state of the receiver. Is this deception?

More recently, several authors have argued that the common requirements that deception benefit a sender and harm a receiver need not hold. Birch (2019) gives a good example. The pied babbler trains young with a purr call emitted during feeding. The call eventually serves to cause approach behavior and begging in the young. But adults subsequently use the same call to summon young away from predators or toward new foraging sites. This looks deceptive but actively benefits the young, potentially at the cost of adult safety when a predator is present. Artiga and Paternotte (2018) propose that to tackle such cases, deception should be defined in terms of function: a deceptive signal is one whose function is to misinform (and where the signal was selected for this purpose). Birch (2019) points out that this account has the same problem with misinformation that Skyrms (2010b) sees – ambiguous signals are also misinformative and so can count as deceptive. Instead, for Birch, deception is strategic exploitation of receivers by senders: the sender exploits a receiver behavior that evolved for the benefits it yields in some other state of the world. It does so by increasing the probability (from the receiver's point of view) that the state obtains. This exploitation need not benefit the sender nor hurt the receiver.

As is probably clear, there is not a clear consensus on what deception looks like in the animal world, just as there is not a clear consensus on content. In the case of deception, though, part of the issue seems to be that we generally ground judgments of what is deceptive in terms of human behavior. It may be that there is no neat, unitary concept underlying these judgments. As such, there may be no single definition of biological deception that captures everything we might want it to do.

As we have seen, there are deep issues related to meaning in animal signals that are not particularly easy to resolve. This discussion ends the part of the Element about signaling. The second part of the Element, as mentioned, focuses on prosociality. Before moving on, it is worth mentioning one current (and fascinating) debate in philosophy of biology that centers around the question of what sort of information is carried in the genome. Does it carry semantic content? Some other kind of content? We will not discuss these questions here for space reasons, but interested readers can look at Maynard-Smith (2000), Godfrey-Smith and Sterelny (2016), Shea (2011), and Bergstrom and Rosvall (2011). And with that, on to part 2.

6 Altruism and the Prisoner's Dilemma

Remember the vampire bats from the introduction? They engage in reciprocal food sharing, where on alternating nights they regurgitate part of their meal to share with a partner. Over time this strategy pays off, since each bat might be the one who is going hungry on a particular evening. Also recall Mitzi and her friend who donated to a kidney chain on her behalf. Thirty-four people lived because they had friends who were willing to literally give away an important bodily organ to save them.

These are examples of *altruism* in the biological sense: one organism forgoes some fitness benefit to increase the fitness of another organism. (This can be contrasted with psychological altruism, which involves an intention to help another.)[57] We see altruistic behavior in many parts of the animal kingdom. Consider the birds who feed young from dawn to dusk, ants who die defending their colony, or the person who drops change in the Salvation Army pot. This sort of behavior poses a puzzle, though. Evolution selects behavior that improves the fitness of an organism. Altruism is defined by the fact that it detriments fitness – how has it evolved?

One prominent approach uses a game called the *prisoner's dilemma* to address this question. This is probably the most famous game in all of game theory and has generated a literature that is truly massive. Furthermore, there are many excellent surveys of this literature available. Another one is not really necessary. This said, no one is allowed to write a book about game theory and biology without discussing the evolution of altruism. So the goal of this section is to give a brief history of work on the evolution of altruism using this framework, focusing on how philosophers have contributed. I will start by describing the model. Then, I discuss the most fundamental lessons on the evolution of altruism that have been developed.[58] I'll move on to a topic that unites the two halves of this Element – how signaling can interact with altruistic behavior. The section will conclude by discussing altruism's dark cousin – spite.

---

[57] There have been many lively debates about how to define altruism. In particular, many think that the common usage implies a psychological intention to help (Joyce, 2007; Okasha, 2013). In addition to being a biological altruist, Mitzi's friend is a psychological altruist as well.

[58] In doing so, I draw on the work of Nowak (2006b), which outlines five key mechanisms by which altruism can evolve.

6.1 The Prisoner's Dilemma

Two prisoners are being asked to rat each other out. If they both stay silent, they each go to prison for one year. If one rats on the other, the rat goes free, while their partner is sent away for five years. If they both rat, though, they are both implicated, and each goes to jail for three years. This scenario induces a payoff structure where each prisoner has a clear preference over outcomes. We can capture these preferences using payoffs in a game.

Consider the payoff table in Figure 6.1. Here we have labeled staying silent as "cooperating" and ratting as "defecting." (This follows tradition, but the reader should note, as will become clear, that cooperation here specifically refers to altruistic behavior rather than mutually beneficial behavior.) Two cooperators get a payoff of 2, while two defectors get a payoff of 1. For this reason, mutual cooperation (staying silent) is preferable to mutual defection (ratting). But notice that despite this, each actor is always incentivized to defect. If their parter is cooperating, defection gets a payoff of 3 (going free), while cooperating only gets a payoff of 2 (one year of jail time). If their partner is defecting, defection gets a payoff of 1 (three years in prison) while cooperating gets a payoff of 0 (five years in prison). This is why the game is a dilemma – despite the benefits of mutual cooperation, there is always a temptation to defect. In fact, the only Nash equilibrium of this game is for both players to defect.

Notice that the cooperation strategy is altruistic in the sense described above. By playing cooperate, a player always raises their partner's payoff while lowering their own. Thus we can use this framework to say more precisely why the evolution of altruism is a puzzle: in standard evolutionary game theoretic models, populations always evolve to a state where all actors defect.[59]

		Player 2	
		Cooperate	Defect
Player 1	Cooperate	2,2	0,3
	Defect	3,0	1,1

Figure 6.1 A payoff table for the prisoner's dilemma. Payoffs for player 1 are listed first.

---

[59] This said, many have argued that the prisoner's dilemma is too simple a model to represent interesting cases of the evolution of altruistic behavior. See, for example, Birch (2012).

6.2 Evolving to Cooperate

The main thrust of the literature mentioned at the beginning of this section is to figure out why, despite the fact that we should expect defection to evolve, vampire bats are vomiting in each others' mouths every night. Clearly something is missing from the model. That something is *correlated interaction*, meaning an increased likelihood of interaction between players who take the same strategies. Imagine if cooperators only met cooperators and defectors only met defectors when playing the prisoner's dilemma. Then cooperators would get a payoff of 2, and defectors 1, and cooperation would straightforwardly be expected to evolve. As it turns out, it is tricky to generate scenarios with perfect correlation of this sort, but there are many mechanisms which generate enough correlation to get altruism off the ground. Let's discuss these in turn.

6.2.1 Kin Selection

Most of the altruism seen in the animal kingdom is the result of kin selection. Hamilton (1964) was the first to identify this mechanism. The basic idea is that many organisms tend to interact disproportionately with kin. These kin are also likely to share genes, including genes that code for behavioral traits like altruism. This means that in many cases the benefits of altruism are falling on other altruists, generating a version of correlated interaction. Under the right conditions, altruism will be selected rather than defection because of this correlation. Many have argued that this process is the starting point of all human ethical behavior: because of the benefits of mutual altruism, animals developed psychological mechanisms leading us to care for our offspring. Eventually these mechanisms were co-opted to extend this care to other kin, and group members who promote the fitness of our offspring and ourselves.[60]

Of course, for this to work, it has to be the case that organisms are somehow able to disproportionately confer the benefits of altruism on their kin rather than unrelated conspecifics. There are two basic ways in which this might happen. The first is through kin recognition, whereby organisms are able to identify their kin (Hamilton, 1964). The second is limited dispersal, whereby kin tend to be closely grouped by dint of being kin, and thus interact more (West et al., 2011).

---

[60] For more on the legacy of Hamilton, and philosophical implications of this legacy, see Birch (2017).

6.2.2 Group Selection

One of the most controversial topics in the evolution of altruism is *group selection*. The idea is that while a group of defecting individuals will tend to have low fitness, a group of altruistic ones – where altruists meet altruists – will do better. So while selection on the individual level should promote the spread of defection, perhaps selection on the group level will promote altruism. There are many confusions that arise in understanding just what it means for selection to happen on the group level, but discussing these is beyond the scope of this Element. (See Okasha, [2006] for a very nice discussion of this and related topics. Birch and Okasha, [2014] weigh in on recent related debates in biology over kin selection and inclusive fitness calculations.)

The controversy is over whether group selection can truly promote the spread of altruism or not. In particular, starting with Maynard-Smith (1964), game theoretic models have been used to show that only very particular conditions will lead to group selection of altruism. Others, including, influentially, Wilson and Sober (1994) and Sober and Wilson (1999), have argued that this evolutionary process is still a plausible one, and that group selection likely played a role in the emergence of human altruism. Some authors, have further argued for the importance of *cultural group selection* in the promotion of prosociality in human groups (Boyd and Richerson, 1985, 1990; Henrich, 2004). The idea is that cultural groups that adopt particularly good prosocial norms and conventions can outperform others that do not. In many cases, there are reasons to think that cultural group selection may be more effective than genetic group selection. This is because humans tend to learn cultural norms and conventions, meaning that there is often a great deal of within-group similarity and between-group difference in cultural practice. In this way, cultural learning creates conditions where group selection is highly likely to occur.[61]

6.2.3 Reciprocity

Trivers (1971) first recognized the potential importance of reciprocity in the promotion of altruism. The idea here is that actors keep track of who has defected and who has cooperated in the past, and use this record to decide whether to cooperate or defect with them in the future. This can generate correlated interaction because cooperators are met with cooperation, and defectors start to meet a disproportionate number of defecting partners.

---

[61] Henrich (2004) argues further that altruism may be selected for if punishment is stabilized by cultural group selection.

There are several forms of reciprocity. *Direct reciprocity* occurs when individuals reciprocate based on their own experience with a partner. This is often modeled via a variation on the prisoner's dilemma called the *iterated prisoner's dilemma*, where actors play the game over multiple rounds.[62] One example of a reciprocating strategy is *tit-for-tat* or TFT. Actors playing this strategy will start by cooperating. Then on successive rounds, they take whatever action their partner did – defecting in response to defect, and cooperating in response to cooperate. This strategy famously won a tournament designed by Robert Axelrod where strategies were put head to head to see which got the highest scores in the iterated prisoner's dilemma (Axelrod and Hamilton, 1981). Furthermore, TFT can out-evolve defection (though the details are a bit more complicated; Trivers, 1971; Axelrod and Hamilton 1981; Nowak (2006a)). Readers interested in learning more about reciprocal altruism in repeated games might look at Binmore (1994, 2005).

Indirect reciprocity occurs when actors condition their behaviors on the interactions of other actors. Alexander (1987) was the first to argue for the importance of this type of reciprocity to the evolution of altruism. In a model of indirect reciprocity, actors might keep track of how often group members have defected on others using some sort of scoring rule,[63] or they might use gossip to transfer information about the behaviors of their group mates. Such models can get quite complicated. Choices must be made, for instance, about who observes what, how gossip is transferred, and how interactions are scored. Plus there are questions, such as, Do you defect against someone who defected against a defectors (Leimar and Hammerstein, 2001)? Or should that count as a boost to reputation? (See Nowak and Sigmund, [2005] for more.) The general take-away, though, is that this sort of reputation tracking can lead to correlated interactions and thus to the spread of altruism.[64]

6.2.4 Network Reciprocity

Network reciprocity has a different flavor from the other forms of reciprocity just mentioned. In those cases, actors who begin to defect are met with defection, and those who start cooperating are met with cooperation. The idea behind network reciprocity is that in many groups, individuals do not freely mix. Instead, actors tend to interact again and again with the same individuals –

---

[62] One typical way to model this is to assume that play will end each round with some probability, α. The size of α determines how long play is likely to last.

[63] See, for example, Nowak and Sigmund (1998).

[64] Smead (2010) looks at a model where actual language for reputation tracking coevolves with indirect reciprocity.

those in spatial proximity or with whom an actor has a relationship of some sort. For various reasons, cooperators might cluster together. This might be because individuals learn or adopt strategies from each other, or because cooperators identify each other and enter stable relationships. This sort of mechanism can be investigated using simulations of network models. Note the connection here to limited dispersal models of kin selection. In both cases, location contributes to the increased correlation between altruists.

In such network models, the dynamics of evolution are often drastically altered. In particular, under many assumptions, altruists can persist in a population or even take over (Alexander, 2007). This usually occurs because altruists are protected from exploitation by the network structure and because those most likely to imitate altruists – those near them in the network – are also those who will subsequently engage in mutually altruistic interactions (Nowak and May, 1992, 1993; Bergstrom and Stark, 1993; Grim, 1995; Alexander, 2007).

6.2.5 Punishment

Another sort of mechanism that might lead to the evolution of cooperation is prosocial punishment. Notice that this mechanism is the only one discussed here that does not work via correlated interaction. The idea is that individuals are punished by group members for social transgression. If so, the benefits of defecting on a cooperator might not be worth the punishment suffered.

In human groups, we see strong tendencies toward moralistic punishment, which suggests that this mechanism may have been important in the emergence of human-level altruism (Fehr and Gächter, 2002; Henrich et al., 2006). There is a bit of a theoretical hiccup, though, which is that punishment is itself usually costly. Thus punishers must decrease their own payoff to improve the social situation of their group. Another question arises then: how might punishment emerge if punishment itself is altruistic? This is sometimes called the *second-order free rider problem* (Heckathorn, 1989).

Various models have been developed to explain how groups might get around this problem (for examples, see Axelrod, 1986; Panchanathan and Boyd, 2004). One possibility is that punishment may not be an effective mechanism for the initial evolution of altruism, but may be able to emerge in altruistic groups to enforce and improve altruistic behavior (Nowak, 2006b).

6.3 Signaling and Altruism

There is another sort of mechanism that can lead to the evolution of altruism in the prisoner's dilemma. I separate this section because this mechanism returns to themes from the first part of the Element. We now look at the relationship between signaling and prosociality.

6.3.1 Secret Handshakes, Greenbeards, and Moral Emotions

Cooperators might be able to identify other cooperators via some sort of signal. This has sometimes been referred to as a "secret handshake" in the economics literature (Robson, 1990). In biology, this kind of secret handshake is sometimes referred to as a *greenbeard*. Imagine a population where only cooperators have green beards and thus can identify each other and choose each other for interactive partners. This can allow cooperation to spread (Dawkins, 1976). As Gardner and West (2010) point out, not all greenbeards require an observable signal of this sort, but we will focus here on the kind of greenbeard that does.[65]

Both secret handshakes and greenbeards of this sort can be modeled as *preplay signals* in games. Before actors play a game – in this case the prisoner's dilemma – they have the opportunity to send a signal. Once the signals are sent, the actors can condition play of the game on the message received.

There is a well-known problem with this sort of preplay signal, though. An individual that knows the secret handshake but always defects can take advantage of the trust of other cooperators to great personal gain. In the biological realm, any mutation that leads to a combination of the greenbeard trait and defecting behavior should spread in a population of greenbearded cooperators. There have been some greenbeards identified in the real world (Gardner and West, 2010), but it may be that this mechanism for cooperation is not widespread because of this vulnerability to invasion (Keller and Ross, 1998).

There are several ways to get around the problem of fakers who know a secret handshake but nonetheless defect (reflecting some lessons from Section 4). One is to have a trait that for whatever reason cannot be faked by defectors. This can be achieved two ways. It might be the case that some signal literally cannot be produced by a defector type, or it might be the case that the signal is more costly for the defector to produce than for a cooperator – or both. Frank (1988) gives an influential argument that moral emotions in humans might play this sort of role. They are associated with prosocial behavior; are, to some degree, observable by other humans; and, he argues, hard to fake. Others have pointed out that guilt – a key moral emotion – is not actually very hard to fake. But

---

[65] For example, the same effect can arise when obligate cooperation is somehow linked to a trait that allows the bearer to benefit from the cooperative behavior in a way that others cannot. Notice that greenbeards should be distinguished from kin recognition mechanisms, mentioned in the last section. Kin recognition mechanisms allow individuals to identify kin. For greenbeards, though, the cooperators can be unrelated outside of the genes in question (Gardner and West, 2010).

since guilt leads to costly apology, these costs may serve to stabilize guilt as a signal that can improve altruism (Okamoto and Matsumura, 2000; O'Connor, 2016; Martinez-Vaquero et al., 2015; Rosenstock and O'Connor, 2018).

Another solution proposed by Santos et al. (2011) is to give actors a large number of signals. Increasing the signals available to actors in their model increases the prevalence of altruism – cooperators can keep out-evolving defectors by selecting new secret handshakes to coordinate their good behavior.[66]

6.4 The Evolution of Spite

In yellow rumped caciques, adolescent males sometimes harass fledglings to such a degree that the fledglings die. In doing so, these adolescent males put themselves in danger of retaliation from adult females and also waste time and energy.

We have now seen how correlation between strategies can lead to the evolution of altruistic behavior. This observation can be flipped, though, to explain the evolution of something a bit darker – spite. Spiteful behavior occurs when individuals pay a cost not to benefit another individual but to harm another individual. After illustrating how kin selection might lead to altruism, Hamilton (1964), and independently George Price, realized that a similar mechanism might lead to the selection of spite (Hamilton, 1970; Price et al., 1970; Hamilton, 1971). In particular, spiteful behavior toward individuals of less than average relatedness might be selected for. Since these individuals tend not to have spiteful genes themselves, harming them can help spite out-compete the population average (Grafen, 1985; Lehmann et al., 2006; West and Gardner, 2010).

It is perhaps clearer to reframe this observation in terms of *anticorrelation*, where, more generally, the harms of spite fall disproportionately on those who are not, themselves, spiteful. This sort of anti-correlation, as in the case of altruism, might be generated multiple ways: through greenbeards (West and Gardner, 2010; Gardner and West, 2010), local interaction (Durrett and Levin, 1994), group selection (Skyrms, 1996), and even due to the anticorrelation generated by the fact that individuals do not interact with themselves (Hamilton, 1970; Smead and Forber, 2013).[67] Smead and Forber (2016) and Forber and Smead (2016) find that, paradoxically, spite can promote altruism by stabilizing mechanisms for recognizing altruistic and selfish types.

---

[66] See also Alexander (2015) for a network model with multiple signals where altruism emerges in the prisoner's dilemma.

[67] Forber and Smead (2014) use a prisoner's delight model to show that, as a result, spite might undermine the evolution of cooperation, even when cooperation yields significant benefits.

Correlation between strategies can lead to the evolution of altruistic behavior, despite its costs. There are many and varied ways this can happen, though. And disagreements still exist over which of these are plausible or more likely to explain real-world altruism. As we will see in the next section, altruism is not the only prosocial game in town, and many of the same mechanisms that promote altruism can help promote the evolution of other sorts of prosocial behavior.

7 Stag Hunting, Bargaining, and the Cultural Evolution of Norms

The prisoner's dilemma game casts a long shadow. Work on the evolution of prosociality has tended to focus heavily on this game (probably in part because of how compelling it is to *solve* a *dilemma*). But there are many aspects of prosociality that are not well captured by the prisoner's dilemma. In this section, we focus on two games that have been used, sometimes together, to shed light on the emergence of prosociality. In particular, many authors have used these games to discuss various aspects of human society, especially the emergence of social contracts and justice. While our discussion in this section veers between the biological and human social realms, we focus on the latter for this reason.[68] We also take some extra time to make clear how game theoretic and evolutionary game theoretic models can capture the emergence of cultural norms in these, and other, games.

In Section 7.1 we look at the stag hunt – a model of scenarios where two individuals have an opportunity to engage in mutually beneficial cooperation, but where doing so is risky. Because humans are so effective at joint action and division of labor, this model is germane to human societies, and especially to social contract theory. This game can be used to model real stag hunting, individuals who clear a field together to graze cattle, who build a town hall they would all like to use, or who create a household together.

The production of joint goods creates opportunities because joint action often involves some synergy – we create more together than separately. It also poses an inherent problem. Actors who work together need to make decisions about (1) who will contribute what share of labor and (2) who will reap what rewards from joint action. In other words, they face a bargaining problem. Bargaining problems are ubiquitous across human society. They occur whenever humans

---

[68] Those who are more focused on the biological realm may wish to read Calcott (2008), who gives a fascinating discussion of various ways that group action among organisms can generate benefit in a stag-hunt-like way. For complementary approaches in philosophy of biology to the evolution of human prosociality, readers might look at the work of Kim Sterelny, including Sterelny (2012) and Tomasello et al. (2012).

engage in joint action, but also in many other situations where resources must be split. In Section 7.2, we look at a bargaining model that has been used both to explain the evolution of justice, and, on the other side of the coin, the emergence of inequity in human societies. We also briefly discuss a model that has been used widely in evolutionary game theory to address resource division among animals. This is actually the first game analyzed using evolutionary game theory – the hawk-dove game.

As I have hinted at, the combination of joint action and resource division is especially important to human functioning. For this reason, in Section 7.3 we see what happens when stag hunting is combined with bargaining in a joint game. The section will conclude by discussing more explicitly a topic that has threaded throughout the Element – the cultural evolution of conventions and norms.

7.1 The Stag Hunt

Two hunters would like to capture as much game as possible. They have two options strategy-wise – to hunt for stag or to hunt for hare. If both hunters go for the stag, they are able to take it down together. If either hunts hare, they will be able to catch the small prey alone. But if one hunts for hare, the remaining hunter cannot catch a stag by themselves. They get no meat if their partner fails to do their part.[69]

The payoff table for this game is shown in Figure 7.1. As we can see, if both players choose stag, they get a payoff of 3. If either chooses hare, they get 2. And hunting stag while your partner hunts hare yields 0. This game, unlike the prisoner's dilemma, has two Nash equilibria. There is a prosocial equilibrium where both players hunt stag, and a loner equilibrium where both hunt hare.

We can take an evolutionary model of the stag hunt to represent a human group who, over time, develop strategies for dealing with cooperative scenarios.

		Player 2	
		Stag	Hare
Player 1	Stag	3,3	0,2
	Hare	2,0	2,2

Figure 7.1 A payoff table for the stag hunt. Payoffs for player 1 are listed first.

---

[69] In *A Discourse on the Origins of Inequality* Rousseau (1984: 27) writes, "If it was a matter of hunting a deer, everyone well realized that he must remain faithful to his post; but if a hare happened to pass within reach of one of them, we cannot doubt that he would have gone off in pursuit of it without scruple."

For this reason, authors like Binmore (1994, 2005) and Skyrms (2004) have taken this game as a simple representation of the social contract. Hare hunting represents an individualistic state of nature, lacking in social trust, and stag hunting the emergence of successful contracts of cooperation and mutual benefit.[70]

It might seem obvious what should emerge in the stag hunt game. There is no dilemma, and the stag equilibrium yields a higher payoff. So we should always see stag hunting, right? Things are not so simple. First of all, both equilibria are ESSs, so either should be stable to invasion. Furthermore, hunting stag is more risky, in this game, than hunting hare.

This extra riskiness means that under typical modeling assumptions stag hunting is less likely to evolve. Hare hunting has larger basins of attraction under the replicator dynamics, because of its more dependable payoffs. Figure 7.2 shows this. For the game in Figure 7.1, a population that is more than two-thirds stag will go to stag hunting, and any proportion less than that will go to hare hunting. Furthermore, in models that incorporate mutation and stochasticity, the prediction is typically that hare hunting, rather than stag hunting will emerge (Kandori et al., 1993). If we imagine a population with random mutations and perturbations, the key observation here is that it is easier to go from stag to hare than from hare to stag.

So in light of this model, how do we explain the widespread success of humans at engaging in joint action? And, likewise, how do we explain wolves who engage in joint hunting? Or bacteria who act together to protect themselves from a hostile environment?

Well, first, we might alter the payoffs a bit. The payoffs for the stag hunt are often constrained to guarantee that hare hunting is *risk dominant*, meaning in

hare   stag

Figure 7.2 The phase diagram for a replicator dynamics model of the stag hunt.

---

[70] The importance of the stag hunt as a model of human interaction does not derive only from straightforward applications of this model to one-shot interactions. Skyrms (2004) points out that the iterated prisoner's dilemma becomes a stag hunt when played between reciprocators and defectors. Bergstrom (2002) also shows how a particular group selection model of the prisoner's dilemma is equivalent to a stag hunt played between the founders of the groups. See Smead (2014b) for a more general treatment of how to model evolving shifts between the stag hunt and the prisoner's dilemma.

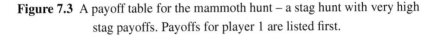

		Player 2	
		Mammoth	Hare
Player 1	Mammoth	100,100	0,2
	Hare	2,0	2,2

Figure 7.3 A payoff table for the mammoth hunt – a stag hunt with very high stag payoffs. Payoffs for player 1 are listed first.

hare   mammoth

Figure 7.4 A phase diagram for the mammoth hunt.

this case that it yields a higher expected payoff given no knowledge of opponent strategy. But we can imagine increasing the payoff to stag hunting until it is so high that stag hunting is risk dominant instead. We can even imagine situations like that in Figure 7.3. We might call this a mammoth hunt. Evolutionary models predict that mammoth is more likely to evolve for this altered model, as is evident in Figure 7.4. Note, though, that hare hunting is still a stable equilibrium. So even for this model, there is a question of how uncooperative populations (states of nature) might evolve to be cooperative (social contract) in the first place.

We can also ask, In situations where stag hunting is moderately beneficial, but risky, what sorts of mechanisms might promote the evolution of cooperation? What allows humans (and other animals) to work together despite the risks? A general message, once again, is that correlated interaction can help stag hunting, just as it helps altruism in the prisoner's dilemma. If stag hunters tend to meet each other with high probability, they do not suffer the costs of their risky prosocial strategy, and so they outperform hare hunters. Again, there are various mechanisms by which this can happen.

7.1.1 Signals and Stag Hunting

Remember in the last section that preplay signaling (secret handshakes or greenbeards or emotions) could play a role in promoting altruism in the prisoner's dilemma. The idea there was that actors could use signals to coordinate on altruism, despite a temptation to defect. Of course, as we saw, because of the nature of the prisoner's dilemma there is always selection for actors who learn how to subvert these systems of communication.

What about preplay signaling in the stag hunt? Skyrms (2002b, 2004) considers a strategy where stag hunters have a secret handshake. They usually hunt

		Player 2	
		Stag	**Hare**
Player 1	**Stag**	3,3	0,2.2
	Hare	2.2,0	2,2

Figure 7.5 A payoff table for the assurance game. Payoffs for player 1 are listed first.

hare, but if they receive a special signal, they hunt stag instead. As he points out, this strategy can invade a population of hare hunters. The stag signalers will always get better payoffs by using their signals to coordinate on beneficial joint action.[71]

This result even holds in a slightly more surprising setting. Consider a version of the stag hunt (sometimes called the *assurance game*) where hare hunters do slightly better against partners who play stag. This is shown in Figure 7.5. In this case, hare hunters are incentivized to mislead potential collaborators as to their strategy, because doing so improves their payoff. For this reason, Aumann (1990) argues that cost-free signals cannot lead to stag hunting in this game. As Skyrms (2002b, 2004) shows, though, the logic of the last paragraph holds even for the assurance game – cheap talk can lead to the emergence of joint action because coordinating stag hunters do so well against each other. Hare hunters are incentivized to mislead but even more incentivized to become stag hunters.

7.1.2 Joint Action with Friends

What about network structure? In the prisoner's dilemma, under the right conditions, such structure can protect clusters of cooperators from invasion by defectors. One might expect something similar for the stag hunt.

Results are slightly complicated. Under some conditions, networks will actually promote hare hunting (Ellison, 1993; Alexander, 2007). The details of network structure, payoffs of the game, updating rules, etc., matter quite a lot, though. Under other conditions, the presence of network interactions will greatly increase the probability that stag hunting emerges (Skyrms, 2004; Alexander, 2007). Consider, for instance, a grid with 1,000 agents who each play the stag hunt with their 8 neighbors. Each round, actors change strategies by switching to whatever strategy their most successful neighbor plays. Although under the replicator dynamics, populations will always go to hare

---

[71] As Santos et al. (2011) and Skyrms (2002b) point out, under some conditions the addition of multiple signals to the stag hunt can make it especially likely that stag hunting evolves by allowing this sort of coordination.

hunting whenever there are fewer than two-thirds stag hunters, Skyrms (2004) finds that in this model, populations with 30 to 40 percent stag hunters tend to end up with at least some stable stag hunting at the end, while for 50 percent or higher, populations are almost guaranteed to end up at all stag. The reason is that stag hunters benefit from repeatedly meeting other stag neighbors, and hare hunters tend to copy their stag neighbors who are reaping the benefits of joint action.

What makes stag versus hare hunting emerge on a network? One thing that makes a difference seems to be how widely agents imitate versus how widely they interact. When agents can observe and imitate the strategies of many neighbors, including those more distant in the network, hare hunters are able to see how well close-knit groups who all play stag do. This leads to the spread of stag hunting. On the other hand, if agents interact more widely, stag hunters are not as protected from the risks of meeting hare hunters, and hare hunting gains a comparative advantage (Skyrms, 2004; Alexander, 2007; Zollman, 2005). As Vanderschraaf and Alexander (2005) point out, another feature that promotes stag hunting is correlated mutation, in the sense that occasionally several agents try stag hunting together. This might result, as they point out, from communication between agents. Their success as a small unit can spread to nearby hare hunters.

There is another way to go about analyzing network interactions, which is to suppose that actors create their own networks. For example, if actors playing stag and hare decide who to interact with based on payoffs, stag hunters will begin to only interact with each other, thus protecting stag hunting (Skyrms and Pemantle, 2000; Skyrms, 2004; Alexander, 2007).

Zollman (2005) looks at what happens if one combines the suggestions from the two last sections. He considers simulations where actors (1) are arranged on a network grid and (2) have two signals they may use to coordinate strategies. His actors practically all evolve to hunt stag, though they use different signals to coordinate in different areas of the network.

7.1.3 Group Selection of Stag Hunting

Binmore (2005) suggests that cultural evolution ought to select for efficient social contracts, by which he has in mind high payoff equilibria in games like the stag hunt. His reasoning appeals to group selection – groups with efficient contracts will expand and give rise to more groups.

Remember that in the prisoner's dilemma, group selection is undermined by individual selection. Even if groups with high levels of altruism do better than those with low, there is always selection within groups for defection.

And for this reason, group selection will only select altruism under restrictive conditions (Maynard-Smith, 1964). But for the stag hunt, a cooperative group is self-enforcing. There is no selection for hare hunting among stag hunters. Authors have translated this argument into models which show that, indeed, group selection can promote the evolution of stag hunting (though this is not a certainty, and the details of how groups evolve matter) (Boyd and Richerson, 1990; Huttegger and Smead, 2011).

This possibility is an especially interesting one in the human sphere because, as noted in Section 6, humans can also undergo cultural group selection. As Boyd and Richerson (1990) and Henrich (2004) point out, this can lead to the selection of efficient ESSs in games like the stag hunt.

Given the often dramatic benefits of joint action for humans, human abilities to communicate, repeated interaction with neighbors, and our tendencies to adopt cultural conventions, stag hunting may, in fact, be easy to get for human cultural groups.

7.2 Bargaining

Two partners in a household must figure out who will contribute how much time to keep the household running. While they share interests in that they wish the household to be successful, both would prefer to do less of the work to keep it that way. Aphids and ants are engaged in a mutualism where ants protect the aphids, who produce a sugar and protein rich nectar for the ants. While the arrangement benefits both of them, there are different levels of nectar the aphids could produce. The ants prefer more and the aphids less. Andrew and Amelia have just cleared a forest where they can now graze cattle. Both benefit from the effort, but now there is a question of who gets to graze how many cattle. Each prefers more for themselves, but too many will ruin the field.

These are scenarios of resource division. In particular, they are all scenarios where individuals benefit by dividing some resource, where there are multiple divisions that might be acceptable to each individual, and where there is conflict over who gets how much. The study of such scenarios in game theory goes back to Nash (1950), who introduced a bargaining problem to represent them. Subsequently, bargaining games have been widely studied in economics and the other social sciences, and applied to cases ranging from household division of labor, to salary negotiations, to international trade deals.

We will consider a relatively simple version of the *Nash demand game* here.[72] Two actors divide a resource of value 10. The actors' strategies consist

---

[72] The original game includes strategies that are demands for any percentage of a resource. Since we will mostly discuss evolutionary models, it is better to look at a game with a finite strategy space.

Player 2

		Low	Med	High
	Low	4,4	4,5	4,6
Player 1	Med	5,4	5,5	0,0
	High	6,4	0,0	0,0

Figure 7.6 A simple version of the Nash demand game. Payoffs for player 1 are listed first.

in demands for some portion of this resource. We will suppose that they can demand either a low, medium, or high amount. Let us constrain these demands so that the medium amount is always for half the resource (5), and the high and low amounts add up to 10. They might be 4 and 6 or 3 and 7. The usual assumption is that if actors make compatible demands, in that they do not over-demand the resource, they get what they ask for. If they make incompatible demands, it is assumed that they failed to achieve a mutually acceptable bargain, and each gets a low payoff called the *disagreement point*. Figure 7.6 shows this game for demands 4, 5, and 6, and a disagreement point of 0.

There are three Nash equilibria of this game. These are the strategy pairings where the actors perfectly divide the resource: Low vs. High, Med vs. Med, and High vs. Low. At these pairings, if either player demands more, they over-demand the resource and get nothing. If either player demands less, they simply get less. Notice that we might alter this simple model to include more possible demands – all the numbers from 1 to 9, for example. In each case, any compatible pairing of demands, 1 and 9, or 3 and 7, will be a Nash equilibrium for the reason just given.

Depending on the number of strategies, then, this game can have many equilibria. If we imagine that actors can make a demand for any percentage of the resource, the game has infinite equilibria, for each possible combination that adds up to the total resource. This poses an *equilibrium selection problem* – what does the game predict vis-à- vis behavior given the surfeit of Nash equilibria? Traditional game theoretic approaches attempt to answer this question via various means (Nash, 1950, 1953; Rubinstein, 1982; Binmore et al., 1986). Here we will discuss how cultural evolutionary models have been used to discuss this problem, focusing on work in philosophy.

7.2.1 The Evolution of Justice

There is something special about the fair demand in a Nash demand game. This demand is the only ESS of the game, despite there being many possible equilibria. This is because the fair demand is the only one that is symmetric in equilibrium – both players can demand the same thing. A group of players demanding Med will always manage to coordinate, no matter who interacts. A group with any other set of strategies will sometimes miscoordinate, Low players will meet Meds and waste the resource, or High players will meet each other and over-demand it. The only truly efficient group is the one we might call completely fair, or completely just.

For this reason, evolutionary models often tend to select fair demands. Under the replicator dynamics, the equilibrium where everyone demands 5 will always have the largest basin of attraction. Furthermore, it will always have a basin of attraction that is greater than 50 percent of the state space (Skyrms, 1996; Alexander and Skyrms, 1999; Skyrms, 2004).[73] In this way, even the most simple evolutionary models seem to help explain justice. Across cultures, humans often have stated fairness norms (Yaari and Bar-Hillel, 1984). And in experimental settings, people pretty much always make fair demands in the Nash demand game.[74] This may be the result of evolved preferences, or tastes, for fairness (Fehr and Schmidt, 1999), or it may be the result of culturally evolved fairness norms (Binmore and Shaked, 2010). In any case, we see in this simple model why evolutionary pressures for either biological or learned fair behaviors might exist.

The simplest models cannot entirely explain fairness norms, though. At the population level, there is another type of equilibrium, sometimes labeled a "fractious" state. At these equilibria, there is some proportion of High demanders, and some Low. When these two types of actors meet, they manage to coordinate, but they mis-coordinate with those playing their own strategies. The overall payoff in such a state is less than in the fair equilibrium for this reason. But these states are stable and are selected by the replicator dynamics with relatively high frequency.

For this reason, philosophers and economists have considered what sorts of models might lead to the emergence of fairness with higher frequency. Again, correlation turns out to be very important. Imagine that strategies tend to meet themselves with high probability. Lows and Highs will do quite poorly. Those making high demands will always reach the disagreement point. Those making low demands will not efficiently divide the resource. Those making medium demands, on the other hand, will always reach equilibrium – perfectly dividing the resource. This kind of correlation can lead to the emergence of fairness (Skyrms, 1996).

---

[73] In addition, Young (1993) shows that the stochastically stable equilibrium (SSE) of such models is the fair one. This is done by looking at a population with mutation or errors and seeing where it spends its time as the probability of error approaches zero.

[74] See Nydegger and Owen (1974), for example.

Alexander and Skyrms (1999) show that local interaction, hashed out as network models, can promote the spread of fairness for this reason. If fair-demanders cluster in a neighborhood, they all reap the benefits of coordinating their expectations symmetrically. This leads to the imitation of fair-demanding. In their simulations of agents on a grid, the emergence of fairness is essentially guaranteed. Alexander (2007) thoroughly investigates how bargaining evolves on networks showing that under many diverse conditions fairness is contagious.[75,76]

7.2.2 The Evolution of Inequity

D'Arms et al. (1998) criticize the claim from Skyrms (1996) that correlation might drive the evolution of fairness. They do not dispute the findings but rather point out that there is another side to the coin. Anti-correlation where unlike strategies tend to meet, will drive the evolution of unfairness rather than fairness. If strategies anti-correlate, Highs and Lows can meet each other and do well, while Meds never meet a type with which they reach equilibrium.

There are different reasons that strategies might anti correlate – for instance, actors might use preplay signaling to match complementary strategies (Skyrms, 1996). One important line of inquiry has considered the role of social categories in facilitating the emergence of inequitable norms by allowing the anti-correlation of bargaining strategies. Consider the following model. There are two sub-groups in a population – let's call them the blues and the yellows. Each individual is aware of the group membership of their interactive partners, and can use this membership to condition their behavior. In other words, each person can learn to develop separate strategies for dealing with blues and yellows.

In replicator dynamics models (and other models) of such populations, the two groups often learn to make fair demands of each other. Importantly, though, such groups also often reach equilibria which are entirely unavailable to populations without subgroups and which involve discrimination and inequity. Consider the following equilibrium. First, every actor makes fair demands of their in-group– yellows demand Med with yellows, and blues with blues. Second, when actors from the two groups meet, blues always demand High and yellows Low. In other words, individuals treat in- and out-group members differently to the disadvantage of one sub-group (Young, 1993; Axtell et al.,

---

[75] Though not all. It's always a bit complicated. See Alexander (2007) for more.

[76] Binmore et al. (1986) and Binmore (2005) use the Nash demand game to give a very influential account of justice, fairness, and social contracts. Reviewing this work goes beyond the scope of this Element.

2000; O'Connor, 2019). This occurs because the sub groups allow anti correlation. In each interaction between groups, there is an extra piece of information available to both actors, "I am a blue, you are a yellow." Overall, the picture that emerges is one where fairness is not the expected evolutionary outcome whenever groups are divided into types (O'Connor, 2019).

These models can help explain why inequity is so persistent between real-world social groups (Axtell et al., 2000; Stewart, 2010). They have been used flexibly to explore a number of scenarios related to categorical inequity: how inequity might arise in networks (Poza et al., 2011; Rubin and O'Connor, 2018), how minority status influences the emergence of discrimination (Bruner, 2017; O'Connor and Bruner, 2019; O'Connor, 2017), how power impacts inequity between social groups (Bruner and O'Connor, 2015; LaCroix and O'Connor, 2017), the impact of division of labor on inequity (Henrich and Boyd, 2008; O'Connor, 2019), intersectional inequity (O'Connor et al., 2019), and how inequity emerges in scenarios of joint action (Henrich and Boyd, 2008).

7.2.3 Hawk-Dove and Resource Division

Some of the earliest evolutionary game theoretic models of resource division did not focus on the Nash demand game, but on a model called *hawk-dove*. Maynard-Smith and Price (1973) present this game as a model of animal conflict (though a version called *chicken* was developed earlier in game theory). Imagine two spiders would like to occupy the same territory. If they are both too aggressive, they will fight to their mutual detriment. If one is aggressive, and the other passive, the aggressive spider will take the territory. If both are passive, they might split the territory, living closer together than either might prefer. Such a scenario can be represented with the payoff table in Figure 7.7. The hawk strategy represents an aggressive one, the dove a passive one.

Notice that we can also interpret this game in the following way: two actors divide a resource of size 4. Two doves will divide it fairly, each getting 2. If a hawk and a dove meet, the hawk will get a high payoff of 3 and the dove 1. If two hawks meet, they will be unable to efficiently divide the resource,

		Player 2	
		Hawk	Dove
Player 1	Hawk	0,0	3,1
	Dove	1,3	2,2

Figure 7.7 A payoff table for hawk-dove. Payoffs for player 1 are listed first.

and will get 0. This interpretation of the game sounds a lot like the Nash demand game, but there are some differences. In hawk-dove, if two actors are engaged in a fair, or passive, interaction, one can always unilaterally change strategies to take advantage of the other. For this reason, there is no fair pure strategy equilibrium. The two pure equilibria are hawk versus dove and dove versus hawk.

There is also a mixed strategy equilibrium where both players get the same expected payoff. This equilibrium, in fact, is the only ESS of the game, which may help explain why damaging animal contests are relatively rare (Sigmund, 2017). In the game shown here, the mixed equilibrium involves both players playing hawk 50 percent of the time and dove 50 percent of the time. When the cost to conflict is high enough, though (i.e., the payoff to hawk vs. hawk is low enough), this equilibrium will involve playing hawk only rarely. The chance that two hawks meet in such a case will be rarer still, meaning that the serious detriments of fighting can be avoided.

But at the ESS hawks sometimes still fight and harm each other, and sometimes doves do not take advantage of their social situation. As in the Nash demand game, mechanisms for anti-correlation can help. Maynard-Smith (1982) introduces the notion of an *uncorrelated asymmetry* – any asymmetry between organisms that can facilitate coordination in a game like hawk-dove. For instance, in the spider example, it may be that one spider is always the current inhabitant of a territory, and one is the intruder. If both spiders adopt a strategy where they play hawk whenever they inhabit and dove whenever they intrude, they will always manage to reach equilibrium. (Maynard-Smith calls this the "bourgeois strategy.") A population playing the mixed strategy ESS can be invaded by a population that pays attention to uncorrelated asymmetries in this way, and thus avoids inefficiencies. The opposite solution – where the intruder plays hawk, and the inhabitant dove – also works to coordinate behavior (Skyrms, 1996). Evolution can select one or the other, though it tends to select the first, possibly because of extra benefits obtained by keeping possession of familiar property.[77]

Hammerstein (1981) expands this analysis to look at cases where there are multiple asymmetries between players. They might be owners or intruders but also might be larger or smaller, for instance. He shows that generically, either can act as an uncorrelated asymmetry, even in cases where one asymmetry

---

[77] Aumann (1974, 1987) introduces another solution, that of the *correlated equilibrium*, where actors pay attention to some external state or device to help them decide which role to play. As Skyrms (1996) points out, a population that pays attention to such a correlating device can invade one that does not, resulting in efficient coordination on roles in hawk-dove.

matters to the payoff table of the game, and the other is entirely irrelevant. This happens because often the benefits of coordination outweigh other payoff considerations. In cases, though, where these coordination benefits are less important, the bourgeois strategy may not be stable. Grafen (1987) points out, for instance, that when territory is necessary for reproduction, desperate organisms may risk everything to oust an owner.

7.3 Bargaining and Joint Production

As mentioned at the beginning of the section, the stag hunt is a particularly good model for joint action. But, of course, when humans want to act together, there are certain questions they have to answer. Who will do what work and how much? Who will reap which rewards? In other words, bargaining, whether implicit or explicit, is necessary for effective joint action. This bargaining can be settled over evolutionary time via the evolution of innate behavior, over cultural time via the evolution of norms, or over the course of one interaction via debate and discussion. But it must be settled. In a literal stag hunt, for example, it is always the case that the hunters must also decide how to divide the meat they have brought in, whether these hunters are humans or wolves. Likewise, mating birds often must first decide whether to mate and then how to divide the labor of raising nestlings (Wagner, 2012).

To address this sort of case, Wagner (2012) analyzes a combined stag hunt/Nash demand game where actors first choose whether to hunt stag or hare. If they hunt stag, they then choose a demand for how much resource they would like and play the Nash demand game. The payoff table for this game is shown in Figure 7.8.[78] Notice that there are four strategies here, Hare, Stag-Low, Stag-Med, and Stag-High. This game again assumes a resource of 10,

		Player 2			
		Stag-Low	Stag-Med	Stag-High	Hare
Player 1	Stag-Low	L,L	L,5	L,H	0,S
	Stag-Med	5,L	5,5	0,0	0,S
	Stag-High	H,L	0,0	0,0	0,S
	Hare	S,0	S,0	S,0	S,S

Figure 7.8 A payoff table for a combined stag hunt/Nash demand game. Payoffs for player 1 are listed first.

---

[78] This is equivalent to a Nash demand game with an outside option. Such games have been analyzed in economics, but here we focus on an evolutionary analysis.

so that the medium demand is for 5. Instead of choosing values for Low and High, this payoff table uses variables L and H where $0 < L < 5 < H < 10$, and where $L + H = 10$. For the Hare payoff we have S (for "solo," since H is taken).

This four-strategy game has four possible Nash equilibria. These are Hare vs. Hare, Stag-Low vs. Stag-High, Stag-Med vs. Stag-Med, and Stag-High vs. Stag-Low. In other words, actors can go it alone or engage in joint production with either fair or unequal divisions of the good produced. Notice that the values of S, L, and H determine just how many of these equilibria will exist in a particular version of the game. In particular, if the value for hare hunting, S, is high enough no one will be willing to get the Low payoff, L. Anyone getting this payoff will instead just switch to solo work. If so, the only remaining equilibria are hare hunting, and stag hunting with an equal division. If $S > 5$, hare hunting is the only remaining equilibrium. Notice that this model displays two sorts of social risk. First there is the risk of the stag hunt – actors who attempt joint action while their partner shirks will get nothing. Second, there is the risk of bargaining. Even if actors agree on joint action, if they do not make compatible demands, they get the disagreement point.

Wagner (2012) shows, surprisingly, that in this combined model, actors are more likely to evolve stag hunting than they are in the stag hunt and more likely to evolve fair bargaining than they are in the Nash demand game. In particular, there are three stable evolutionary outcomes of this game under the replicator dynamics – all Hare, all Stag-Med, or some Stag-Low and some Stag-High (this is the "fractious" equilibrium that we saw in the Nash demand game). As the payoff for Hare increases, it becomes more and more likely that the fair outcome emerges than the fractious one. Once the payoff for Hare is high enough, it disrupts the stability of the fractious outcome altogether. And, for many values of S, the combined basins of attraction for stag hunting are larger than in the stag hunt alone. There seems to be a positive message here for prosociality – the dynamics of joint production and division of labor tend to yield good outcomes.

Once again, though, anti-correlation can change this picture. Bruner and O'Connor (2015) look at the stag hunt/Nash demand game described here, but again consider models with multiple groups or social categories. As they point out, inequitable divisions of resource commonly emerge between these groups, because, once again, group structure allows actors to break symmetry. One group becomes the group that gets low, and the other the group that gets high. As they also show, this kind of inequity can lead to a dissolution of joint action (or the social contract, if we like). A group that expects to get too little from a joint project may be unwilling to take on the risks of collaboration, and decide to hunt hare instead. This happens even when joint action provides benefits to both sides, because it is also risky.[79]

7.4 Norms, Conventions, and Evolution

This section has now considered several models that have been taken to inform the cultural evolution of conventional, and sometimes normative behavior. Over the years, philosophers have provided many accounts of what conventions are. We will not overview these here. But, generally, conventions are patterns of social behavior that solve some sort of social problem but that might have been otherwise. Lewis (1969) influentially used games as a way to define conventions. For him, they are equilibria in games with multiple equilibria, along with certain beliefs for the actors.[80] On this sort of account, it is easy to see how the work described in this section, and earlier in the Element, can be taken to inform the cultural emergence of human conventions. Whenever we have a model where actors might evolve to various equilibria, and where we can show that learning processes tend to lead actors to stable, equilibrium patterns, we have a model that might represent the emergence of human convention. This could be a linguistic convention as in Section 3, a convention of adhering to a social contract, a convention for fair treatment, or a convention of unequal resource division. The suggestion is not that all conventions are equilibria in games but that evolutionary game theoretic models provide simple representations of many sorts of emerged social conventions. (Readers interested in more on game theory and convention might look at the work of Peter Vanderscraaf, including Vanderschraaf [1995, 1998a, 1998b], and of economist Ken Binmore [1994, 2005].)

Social norms are also guides to behavior, but of a different sort. The key factor that makes something a norm is the judgment that actors *ought* to follow it. Some norms are also conventions. For instance, it is conventional to stand on the right and walk on the left of escalators in London. But many underground riders also feel that everyone ought to conform to this behavior. Anyone who does not risks social disapproval. Other norms are not conventions.[81] But, in general, we might ask whether evolutionary game theoretic models can tell us

---

[79] Cochran and O'Connor (2019) also show that in similar models where actors explicitly divide both time spent and resources earned on a project, inequity is always highly likely between sub-groups.

[80] I am being handwavy here. In fact, Lewis limits the set of games where one might have a convention to those where at equilibrium actors cannot change behaviors and benefit any other actor. And he gives detailed conditions for common knowledge needed to sustain a convention on his account.

[81] For instance, Bicchieri (2005) describes norms that are not conventions because individuals studiously avoid creating the situations in which the normative behavior would be required.

anything about norms. This is especially so given that these models often do tell us about normatively proscribed prosocial behaviors like altruism, cooperation, and fairness.

An evolved equilibrium in a model fails to capture the ought inherent in norms. Actors adhere to the behavior because it is personally beneficial, not because they ought to. Bicchieri (2005) uses standard game theory to give an influential account of how norms work. For her, norms can transform one game into another one, by making it the case that individuals prefer certain behavior on the understanding that it is expected of them, or on the belief that others will adhere to it, or punish them for non-compliance.[82]

Evolutionary models can also be used to elucidate norms. As Harms and Skyrms (2008) point out, norms are tightly connected to retribution and punishment. Individuals are often willing to punish norm violators, and it seems that humans have an evolved tendency to do so. As such, understanding the evolution of punishing in strategic scenarios may help us understand our normative behavior. As mentioned in Section 6, costly punishment poses a second-order free rider problem – why should people punish others altruistically? Nonetheless, various extant models argue that altruistic punishment can evolve (Boyd et al., 2003). Furthermore, ostracizing behaviors like avoiding defectors, or defecting against them in the prisoner's dilemma, are arguably types of punishment that actually lead to fitness benefits (Harms and Skyrms, 2008). So evolutionary game theoretic models can shed light on this aspect of normativity.

In this section we looked at models that have been widely applied to the cultural evolution of human conventions, especially those related to mutually beneficial cooperation and resource division. As we saw, these models help explain the emergence of social contracts, norms of fairness, and the ubiquity of unfairness. Some themes from Section 6 were repeated here, including the importance of signaling and of interaction structure to the emergence of prosocial behaviors.

8 Epilogue: Models, Methods, and the Philosophy of Biology

One theme that has tracked through this Element regards methodological critique. By way of closing out the Element, I would like to take a little space to discuss some more explicit methodological critiques offered by philosophers of biology.

---

[82] Skyrms and Zollman (2010) consider a suggestion from Bicchieri that individuals often activate "scripts" to decide which norm to follow using an evolutionary model. They consider what happens if individuals treat various bargaining games as equivalent or activated by the same script.

As Huttegger and Zollman (2013) point out, there is something strange about the fact that evolutionary biology, a discipline explicitly devoted to understanding the dynamics of evolution, has relied so heavily on ESS methodology. Many analyses of evolutionary systems have proceeded by finding ESSs, without further work. Of course, there are reasons for this. ESS methodology is relatively easy and straightforward. Furthermore, it is generalizable across a wide class of dynamics. One does not need to painstakingly analyze many variants on evolutionary rules if it is possible to identify the equilibria that are stable across them.

Despite these advantages, Huttegger and Zollman criticize ESS methodology, drawing on some of the work we have seen throughout this Element. They make several important points. First, ESSs do not exhaust the stable states that emerge under common evolutionary dynamics. Partial pooling equilibria in the signaling game, for example, can evolve but are not ESSs. Second, identifying ESSs does not say much about basins of attraction. As we have seen, one ESS might be more likely to emerge than another, for instance, in the stag hunt. And, more pressing, sometimes non-ESS outcomes are more likely to emerge than ESS outcomes in similar models, as in the case of the hybrid equilibrium. In other words, while ESS analysis is an important tool in the evolutionary game theory kit, it should be combined with more detailed dynamical analyses.[83]

This is not the only work in philosophy criticizing simplifying assumptions that surround ESS methodology. For example, Rubin (2016) criticizes the common assumption, in evolutionary models, that we can ignore genetics. Evolutionary game theoretic models usually act as if phenotypic traits themselves are straightforwardly inherited from generation to generation. However, including explicit genetics can often put a wrench in the gears, which Rubin shows by analyzing genetic models of several games. She further shows, expanding on previous work by biologists, that in explicitly genetic models, ESSs may be unattainable. When this is the case, ESSs are obviously not appropriate predictors of evolutionary outcomes.

Likewise, philosophers have criticized influential accounts from Harley (1981) and Maynard-Smith (1982), who argue that learning strategies that do not learn ESSs should never evolve. As the argument goes, otherwise, they should be replaced by those that do learn ESSs, and so improve the payoffs of the organism in question. But, as Smead (2012) points out, since learning is usually cognitively costly, why bother learning an ESS? Why not just adopt it

---

[83] Wagner (2013a) goes even further by showing how non-Nash outcomes (including chaos) emerge in evolutionary models.

simpliciter?[84] The answer is that learning is only helpful in variable environments where organisms might want to switch ESS behavior.[85] But, as O'Connor (2017) points out, Harley and Maynard-Smith ignore the fact that speed and precision necessarily trade off for many learning strategies. This means that in variable environments, where speed matters, organisms can often do better with quick-learning strategies that get close enough to ESS play rather than high-precision strategies that learn ESSs eventually.[86]

Beyond critiques of ESS methodology, philosophers of biology, and philosophers of science generally, have asked wider questions: What sorts of knowledge or understanding do we derive from evolutionary models? Do these simplified models explain? What are the pitfalls of using models to inform evolutionary biology and cultural evolution? There is a large literature in philosophy of modeling that seeks to answer these questions. I will not review it here but will draw on a few central observations relevant to the work presented in this Element. A standard worry is that simplified evolutionary models can mislead. Because model systems are not the real-world systems they correspond to, sometimes observations of these systems will not match the relevant features of the world. This sometimes leads to mistaken conclusions. And, given this, academics do not always treat conclusions drawn from models with the proper level of skepticism.

In any particular investigation, the details of the case will determine what sorts of epistemic roles the models are justified in playing. These roles are many and varied.[87] For instance, simple evolutionary models can generate hypotheses, direct empirical research, turn over impossibility claims, and generate counterfactual information. Perhaps the most controversial roles they play are ones directed at our immediate understanding of the natural world. Do evolutionary signaling models, for instance, actually tell us why peacocks have large tails? In explorations of this sort, an important tool is *robustness testing* (Levins, 1966; Weisberg, 2012). This involves changing various aspects of a model to see whether results are affected. If not, we might increase our confidence that the results are expected in the real world or that some causes will reliably produce some output. There are debates about how and, indeed,

---

[84] See also Smead and Zollman (2009). Maynard-Smith (1982) was aware of this problem.

[85] See Godfrey-Smith (1998) for a more general treatment of this idea that learning, or cognition generally, is only useful in variable or complex environments.

[86] See Smead (2014c) for further models pressing on the analyses of Harley and Maynard-Smith. And see Zollman and Smead (2010) for an evolutionary game theoretic model that emphasizes the importance of short-term learning.

[87] See, for instance, Downes (1992) and O'Connor and Weatherall (2016).

whether robustness analysis works, but details are beyond the purview of this discussion.[88]

There is a very relevant sub-debate though, about the role of basins of attraction in explanation. Analyzing the size of basins of attraction can be thought of as a species of robustness testing. We vary population starting points across models to see what outcomes are particularly stable in different scenarios. Throughout the Element, we have seen inferences that large basins of attraction imply that some equilibrium is evolutionarily important or likely to emerge. But this might lead us wrong if we do not take into account the likelihood that real populations start at various states. In other words, dynamical methods themselves have issues, though ones that are different from ESS methods. There are no perfect modeling methods that will yield clear, easy truths. Instead, models are tools that can improve inference on some topic but should never be used uncritically.

What we see across all the work described in this epilogue is a willingness to step outside the dominant methodological paradigms to assess their applicability. Of course, biologists do this all the time. But there might be something special about the relationship between philosophy and biology that facilitates this sort of work. Philosophers of science are generally trained to look at science with a critical eye. If this is combined with genuine engagement within an area of science, as in the case of evolutionary game theory, it is unsurprising that the outcome is critique and development of the methods in question. In the best cases, this has led to an improvement of the science at hand. Notably, in the literature surveyed here, this improvement has only been possible because philosophers have gained mastery over the tools of evolutionary game theory. In other words, these are not critiques from without but critiques from within by sets of academics with interdisciplinary training. Ultimately, good modelers in biology and philosophy of biology will strive to fulfill this double role – using modeling tools to understand the world, while also sharpening these tools to better serve their epistemic uses.

---

[88] One nice account by Schupbach (2016) argues that robustness checks rule out competing explanations for some result.

Bibliography

Alexander, J. McKenzie (2007). *The structural evolution of morality*. Cambridge, UK: Cambridge University Press.

Alexander, J. McKenzie (2014). "Learning to signal in a dynamic world." *The British Journal for the Philosophy of Science*, *65*(4), 797–820.

Alexander, J. McKenzie (2015). "Cheap talk, reinforcement learning, and the emergence of cooperation." *Philosophy of Science*, *82*(5), 969–82.

Alexander, Jason, and Brian Skyrms (1999). "Bargaining with neighbors: Is justice contagious?" *The Journal of Philosophy*, *96*(11), 588–98.

Alexander, Jason McKenzie, Brian Skyrms, and Sandy L. Zabell (2012). "Inventing new signals." *Dynamic Games and Applications*, *2*(1), 129–45.

Alexander, Richard (1987). *The Biology of Moral Systems*. Abingdon-on-Thames, UK: Routledge.

Argiento, Raffaele, Robin Pemantle, Brian Skyrms, and Stanislav Volkov (2009). "Learning to signal: Analysis of a micro-level reinforcement model." *Stochastic Processes and Their Applications*, *119*(2), 373–90.

Artiga, Marc, and Cédric Paternotte (2018). "Deception: A functional account." *Philosophical Studies*, *175*(3), 579–600.

Aumann, Robert J. (1974). "Subjectivity and correlation in randomized strategies." *Journal of Mathematical Economics*, *1*(1), 67–96.

Aumann, Robert J. (1987). "Correlated equilibrium as an expression of Bayesian rationality." *Econometrica: Journal of the Econometric Society*, *55*(1) 1–18.

Axelrod, Robert (1986). "An evolutionary approach to norms." *American Political Science Review*, *80*(4), 1095–1111.

Axelrod, Robert, and William Donald Hamilton (1981). "The evolution of cooperation." *Science*, *211*(4489), 1390–96.

Axtell, Robert, Joshua M. Epstein, and H. Peyton Young (2000). "The emergence of classes in a multi-agent bargaining model." In *Social Dynamics*. Ed. Steven N. Durlauf and Peyton Young. MIT Press, Cambridge, MA.

Barrett, Jeffrey, Brian Skyrms, and Calvin Cochran (2018). "Hierarchical models for the evolution of compositional language." Unpublished.

Barrett, Jeffrey, and Kevin J. S. Zollman (2009). "The role of forgetting in the evolution and learning of language." *Journal of Experimental and Theoretical Artificial Intelligence*, *21*(4), 293–309.

Barrett, Jeffrey A. (2006). "Numerical simulations of the Lewis signaling game: Learning strategies, pooling equilibria, and the evolution of grammar." Unpublished.

Barrett, Jeffrey A. (2007). "Dynamic partitioning and the conventionality of kinds." *Philosophy of Science, 74*(4), 527–46.

Barrett, Jeffrey A. (2009). "The evolution of coding in signaling games." *Theory and Decision, 67*(2), 223–37.

Barrett, Jeffrey A. (2010). "Faithful description and the incommensurability of evolved languages." *Philosophical Studies, 147*(1), 123.

Barrett, Jeffrey A. (2013a). "The evolution of simple rule-following." *Biological Theory, 8*(2), 142–50.

Barrett, Jeffrey A. (2013b). "On the coevolution of basic arithmetic language and knowledge." *Erkenntnis, 78*(5), 1025–36.

Barrett, Jeffrey A. (2014a). "The evolution, appropriation, and composition of rules." *Synthese, 195*(2), 623–636.

Barrett, Jeffrey A. (2014b). "On the coevolution of theory and language and the nature of successful inquiry." *Erkenntnis, 79*(4), 821–34.

Barrett, Jeffrey A. (2014c). "Rule-following and the evolution of basic concepts." *Philosophy of Science, 81*(5), 829–39.

Barrett, Jeffrey A. (2016). "On the evolution of truth." *Erkenntnis, 81*(6), 1323–32.

Barrett, Jeffrey A. (2017). "Truth and probability in evolutionary games." *Journal of Experimental and Theoretical Artificial Intelligence, 29*(1), 219–25.

Barrett, Jeffrey A., and Brian Skyrms (2017). "Self-assembling games." *British Journal for the Philosophy of Science, 68*(2), 329–353.

Bergstrom, Carl T., and Michael Lachmann (1997). "Signalling among relatives. I. Is costly signalling too costly?" *Philosophical Transactions of the Royal Society of London B: Biological Sciences, 352*(1353), 609–17.

Bergstrom, Carl T., and Michael Lachmann (1998). "Signaling among relatives. III. Talk is cheap." *Proceedings of the National Academy of Sciences, 95*(9), 5100–5105.

Bergstrom, Carl T., and Martin Rosvall (2011). "The transmission sense of information." *Biology and Philosophy, 26*(2), 159–76.

Bergstrom, Theodore C. (2002). "Evolution of social behavior: Individual and group selection." *Journal of Economic Perspectives, 16*(2), 67–88.

Bergstrom, Theodore C., and Oded Stark (1993). "How altruism can prevail in an evolutionary environment." *American Economic Review, 83*(2), 149–55.

Bicchieri, Cristina (2005). *The grammar of society: The nature and dynamics of social norms*. Cambridge: Cambridge University Press.

Binmore, Ken (2005). *Natural justice*. Oxford: Oxford University Press.

Binmore, Ken, Ariel Rubinstein, and Asher Wolinsky (1986). "The Nash bargaining solution in economic modelling." *RAND Journal of Economics, 17*(2), 176–88.

Binmore, Kenneth, and Avner Shaked (2010). "Experimental economics: Where next? Rejoinder." *Journal of Economic Behavior and Organization*, *73*(1), 120–21.

Binmore, Kenneth George (1994). *Game theory and the social contract I: Playing fair*. Cambridge, MA: MIT Press.

Birch, Jonathan (2012). "Collective action in the fraternal transitions." *Biology and Philosophy*, *27*(3), 363–80.

Birch, Jonathan (2014). "Propositional content in signalling systems." *Philosophical Studies*, *171*(3), 493–512.

Birch, Jonathan (2017). *The philosophy of social evolution*. Oxford: Oxford University Press.

Birch, J. (2019). Altruistic deception. *Studies in History and Philosophy of Science Part C: Studies in History and Philosophy of Biological and Biomedical Sciences, 74*, 27–33

Birch, Jonathan, and Samir Okasha (2014). "Kin selection and its critics." *BioScience*, *65*(1), 22–32.

Blume, Andreas, and Oliver Board (2013). "Language barriers." *Econometrica*, *81*(2), 781–812.

Blume, Andreas, Douglas V. DeJong, Yong-Gwan Kim, and Geoffrey B. Sprinkle (1998). "Experimental evidence on the evolution of meaning of messages in sender-receiver games." *American Economic Review*, *88*(5), 1323–40.

Blume, Andreas, Yong-Gwan Kim, and Joel Sobel (1993). "Evolutionary stability in games of communication." *Games and Economic Behavior*, *5*(4), 547–75.

Boyd, Robert, Herbert Gintis, Samuel Bowles, and Peter J. Richerson (2003). "The evolution of altruistic punishment." *Proceedings of the National Academy of Sciences*, *100*(6), 3531–35.

Boyd, Robert, and Peter J. Richerson (1985). *Culture and the evolutionary process*. Chicago, IL: University of Chicago Press.

Boyd, Robert, and Peter J. Richerson (1990). "Group selection among alternative evolutionarily stable strategies." *Journal of Theoretical Biology*, *145*(3), 331–42.

Brown, George W. (1951). "Iterative solution of games by fictitious play." *Activity Analysis of Production and Aladdress*, *13*(1), 374–76.

Bruner, Justin, Cailin O'Connor, Hannah Rubin, and Simon M. Huttegger (2018). "David Lewis in the lab: Experimental results on the emergence of meaning." *Synthese*, *195*(2), 603–21.

Bruner, Justin, and Hannah Rubin (2018). "Inclusive fitness and the problem of honest communication." *British Journal for the Philosophy of Science*, https://doi.org/10.1093/bjps/axy028.

Bruner, Justin P. (2017). "Minority (dis)advantage in population games." *Synthese*, *196*(1), 413–27.

Bruner, Justin P., and Cailin O'Connor (2015). "Power, bargaining, and collaboration." In *Scientific collaboration and collective knowledge*. Ed. Conor Mayo-Wilson Thomas Boyer and Michael Weisberg. Oxford: Oxford University Press.

Calcott, Brett (2008). "The other cooperation problem: Generating benefit." *Biology and Philosophy*, *23*(2), 179–203.

Calcott, Brett (2014). "The Creation and reuse of information in gene regulatory networks." *Philosophy of Science*, *81*(5), 879–90.

Cao, Rosa (2012). "A teleosemantic approach to information in the brain." *Biology and Philosophy*, *27*(1), 49–71.

Caro, T. M. (1986). "The functions of stotting in Thomson's gazelles: Some tests of the predictions." *Animal Behaviour*, *34*(3), 663–84.

Caro, T. M., Leslie Lombardo, A. W. Goldizen, and Marcella Kelly (1995). "Tail-flagging and other antipredator signals in white-tailed deer: New data and synthesis." *Behavioral Ecology*, *6*(4), 442–50.

Cochran, Calvin, and Cailin O'Connor (2019). "Inequity and inequality in the emergence of conventions." *18*(3), 264–281.

Crane, Tim (2015). *The mechanical mind: A philosophical introduction to minds, machines and mental representation.* Abingdon-on-Thames, UK: Routledge.

Crawford, Vincent P., and Joel Sobel (1982). "Strategic information transmission." *Econometrica: Journal of the Econometric Society*, *50*(6), 1431–51.

D'Arms, Justin, Robert Batterman, and Krzyzstof Górny (1998). "Game theoretic explanations and the evolution of justice." *Philosophy of Science*, *65*(1), 76–102.

Dawkins, Richard (1976). *Theory of games and economic behavior.* New York: Oxford University Press.

Donaldson, Matina C., Michael Lachmann, and Carl T. Bergstrom (2007). "The evolution of functionally referential meaning in a structured world." *Journal of Theoretical Biology*, *246*(2), 225–33.

Downes, Stephen M. (1992). "The importance of models in theorizing: A deflationary semantic view." In *PSA: Proceedings of the biennial meeting of the Philosophy of Science Association*, 142–53.

Dretske, Fred (1981). *Knowledge and the flow of information.* Cambridge, MA: MIT Press.

Durrett, Richard, and Simon Levin (1994). "The importance of being discrete (and spatial)." *Theoretical Population Biology*, *46*(3), 363–94.

Ellison, Glenn (1993). "Learning, local interaction, and coordination." *Econometrica: Journal of the Econometric Society*, *61*(5), 1047–71.

Fallis, Don, and Peter J. Lewis (2019). "Toward a formal analysis of deceptive signaling." *Synthese*, *196*(6), 2279–2303.

Fehr, Ernst, and Simon Gächter (2002). "Altruistic punishment in humans." *Nature*, *415*(6868), 137–140.

Fehr, Ernst, and Klaus M. Schmidt (1999). "A theory of fairness, competition, and cooperation." *Quarterly Journal of Economics*, *114*(3), 817–68.

Fitzgibbon, Claire D., and John H. Fanshawe (1988). "Stotting in Thomson's gazelles: An honest signal of condition." *Behavioral Ecology and Sociobiology*, *23*(2), 69–74.

Forber, Patrick, and Rory Smead (2014). "An evolutionary paradox for prosocial behavior." *Journal of Philosophy*, *111*(3), 151–66.

Forber, Patrick, and Rory Smead (2016). "The evolution of spite, recognition, and morality." *Philosophy of Science*, *83*(5), 884–96.

Frank, Robert H. (1988). *Passions within reason: The strategic role of the emotions*. New York: W. W. Norton.

Franke, Michael (2012). "On assertoric and directive signals and the evolution of dynamic meaning." *International Review of Pragmatics*, *4*(2), 232–60.

Franke, Michael (2013a). "An adaptationist criterion for signal meaning." In *The dynamic, inquisitive, and visionary life of φ*, eds: Maria Aloni, Michael Franke, and Floris Roelofsen, 96–104.

Franke, Michael (2013b). "Game theoretic pragmatics." *Philosophy Compass*, *8*(3), 269–84.

Franke, Michael (2014). "Creative compositionality from reinforcement learning in signaling games." In *Evolution of Language: Proceedings of the 10th International Conference (EVOLANG10)*, 82–89.

Franke, Michael (2016). "The evolution of compositionality in signaling games." *Journal of Logic, Language, and Information*, *25*(3–4), 355–77.

Franke, Michael, and José Pedro Correia (2017). "Vagueness and imprecise imitation in signalling games." *British Journal for the Philosophy of Science* *69*(4), 1037–1067.

Franke, Michael, et al. (2009). *Signal to act: Game theory in pragmatics*. Institute for Logic, Language, and Computation.

Franke, Michael, Gerhard Jäger, and Robert Van Rooij (2010). "Vagueness, signaling and bounded rationality." In *JSAI international symposium on artificial intelligence*, 45–59.

Fudenberg, Drew, and Jean Tirole (1991). *Game theory*. Cambridge, MA: MIT Press.

Gardner, A., and S. West (2010). "Greenbeards." *Evolution*, *61*(1), 25–38.

Gibbons, Robert (1992). *Game theory for applied economists*. Princeton, NJ: Princeton University Press.

Gintis, Herbert (2000). *Game theory evolving: A problem-centered introduction to modeling strategic behavior*. Princeton, NJ: Princeton University Press.

Godfrey-Smith, Peter (1989). "Misinformation." *Canadian Journal of Philo sophy*, *19*(4), 533–50.

Godfrey-Smith, Peter (1998). *Complexity and the function of mind in nature*. Cambridge: Cambridge University Press.

Godfrey-Smith, Peter (2011). "Signals: Evolution, learning, and information, by Brian Skyrms." *120*, 1288–1297.

Godfrey-Smith, Peter (2013). "Signals, icons, and beliefs." In *Millikan and her critics*, eds: Dan Ryder, Justine Kingsbury, Kenneth Williford, Wiley-Blackwell Malden, MA, 41–62.

Godfrey-Smith, Peter (2014). "Sender-receiver systems within and between organisms." *Philosophy of Science*, *81*(5), 866–78.

Godfrey-Smith, Peter, and Manolo Martínez (2013). "Communication and common interest." *PLOS Computation Biology*, *9*(11), e1003282.

Godfrey-Smith, Peter, and Kim Sterelny (2016). "Biological information." In *The Stanford encyclopedia of philosophy*. Ed. Edward N. Zalta. https://plato.stanford.edu/entries/information-biological/

Grafen, Alan (1982). "How not to measure inclusive fitness." *Nature*, *298*(5873), 425–426.

Grafen, Alan (1985). "A geometric view of relatedness." *Oxford Surveys in Evolutionary Biology*, *2*(2), 28–89.

Grafen, Alan (1987). "The logic of divisively asymmetric contests: respect for ownership and the desperado effect." *Animal Behaviour*, *35*(2), 462–67.

Grafen, Alan (1990). "Biological signals as handicaps." *Journal of Theoretical Biology*, *144*(4), 517–46.

Grice, H. Paul (1957). "Meaning." *The Philosophical Review*, *66*(3), 377–88.

Grice, H. Paul (1991). *Studies in the way of words*. Cambridge, MA: Harvard University Press.

Grim, Patrick (1995). "The greater generosity of the spatialized prisoner's dilemma." *Journal of Theoretical Biology*, *173*(4), 353–59.

Grose, Jonathan (2011). "Modelling and the fall and rise of the handicap principle." *Biology and Philosophy*, *26*(5), 677–96.

Güth, Werner, Rolf Schmittberger, and Bernd Schwarze (1982). "An experimental analysis of ultimatum bargaining." *Journal of Economic Behavior and Organization*, *3*(4), 367–88.

Hamilton, William D. (1964). "The genetical evolution of social behaviour. II." *Journal of Theoretical Biology*, *7*(1), 17–52.

Hamilton, William D. (1970). "Selfish and spiteful behaviour in an evolutionary model." *Nature*, *228*(5277), 1218–1222.

Hamilton, William D. (1971). "Selection of selfish and altruistic behavior in some extreme models." *Man and Beast: Comparative Social Bahavior*, eds: JF Eisenberg and WS Dillon, Smithsonian Press, Washington DC 57–91.

Hammerstein, Peter (1981). "The role of asymmetries in animal contests." *Animal behaviour*, *29*(1), 193–205.

Harley, Calvin B. (1981). "Learning the evolutionarily stable strategy." *Journal of theoretical biology*, *89*(4), 611–33.

Harms, William, and Brian Skyrms (2008). "Evolution of moral norms." In *Oxford handbook in the philosophy of biology*. Ed. Michael Ruse. Oxford: Oxford University Press.

Harms, William F. (2004a). *Information and meaning in evolutionary processes*. Cambridge: Cambridge University Press.

Harms, William F. (2004b). "Primitive content, translation, and the emergence of meaning in animal communication." In *Evolution of communication systems: A comparative approach*, Eds. GB Müller, T Pradeu, K Schäfer The MIT Press, Cambridge MA. 31–48.

Heckathorn, Douglas D. (1989). "Collective action and the second-order free-rider problem." *Rationality and Society*, *1*(1), 78–100.

Henrich, Joseph (2004). "Cultural group selection, coevolutionary processes and large-scale cooperation." *Journal of Economic Behavior and Organization*, *53*(1), 3–35.

Henrich, Joseph, and Robert Boyd (2008). "Division of labor, economic specialization, and the evolution of social stratification." *Current Anthropology*, *49*(4), 715–24.

Henrich, Joseph, Richard McElreath, Abigail Barr, Jean Ensminger, Clark Barrett, Alexander Bolyanatz, Juan Camilo Cardenas, Michael Gurven, Edwins Gwako, Natalie Henrich et al. (2006). "Costly punishment across human societies." *Science*, *312*(5781), 1767–70.

Herrnstein, Richard J. (1970). "On the law of effect." *Journal of the Experimental Analysis of Behavior*, *13*(2), 243–66.

Hofbauer, Josef, and Simon M. Huttegger (2008). "Feasibility of communication in binary signaling games." *Journal of Theoretical Biology*, *254*(4), 843–49.

Hopkins, Ed (2002). "Two competing models of how people learn in games." *Econometrica*, *70*(6), 2141–66.

Huttegger, Simon, Brian Skyrms, Pierre Tarrès, and Elliott Wagner (2014). "Some dynamics of signaling games." *Proceedings of the National Academy of Sciences*, *111*(Suppl. 3), 10873–80.

Huttegger, Simon M. (2007a). "Evolution and the explanation of meaning." *Philosophy of Science, 74*(1), 1–27.

Huttegger, Simon M. (2007b). "Evolutionary explanations of indicatives and imperatives." *Erkenntnis, 66*(3), 409–36.

Huttegger, Simon M. (2007c). "Robustness in signaling games." *Philosophy of Science, 74*(5), 839–47.

Huttegger, Simon M., Justin P. Bruner, and Kevin J. S. Zollman (2015). "The handicap principle is an artifact." *Philosophy of Science, 82*(5), 997–1009.

Huttegger, Simon M., Brian Skyrms, Rory Smead, and Kevin J. S. Zollman (2010). "Evolutionary dynamics of Lewis signaling games: Signaling systems vs. partial pooling." *Synthese, 172*(1), 177–91.

Huttegger, Simon M., Brian Skyrms, and Kevin J. S. Zollman (2014). "Probe and adjust in information transfer games." *Erkenntnis, 79*(4), 835–53.

Huttegger, Simon M., and Rory Smead (2011). "Efficient social contracts and group selection." *Biology and Philosophy, 26*(4), 517–31.

Huttegger, Simon M., and Kevin J. S. Zollman (2010). "Dynamic stability and basins of attraction in the Sir Philip Sidney game." *Proceedings of the Royal Society of London B: Biological Sciences, 277*(1689), 1915–22.

Huttegger, Simon M., and Kevin J. S. Zollman (2011a). "Evolution, dynamics, and rationality: The limits of ESS methodology." *Evolution and Rationality: Decisions, Co-operation and Strategic Behaviour*, Eds. S. Okasha and K Binmore, Cambridge University Press, Cambridge, MA *67* 67–83.

Huttegger, Simon M., and Kevin J. S. Zollman (2011b). "Signaling games." In *Language, games, and evolution*, 160–76. Berlin: Springer.

Huttegger, Simon M., and Kevin J. S. Zollman (2013). "Methodology in biological game theory." *British Journal for the Philosophy of Science, 64*(3), 637–658.

Huttegger, Simon M., and Kevin J. S. Zollman (2016). "The robustness of hybrid equilibria in costly signaling games." *Dynamic Games and Applications, 6*(3), 347–58.

Jäger, Gerhard (2007). "The evolution of convex categories." *Linguistics and Philosophy, 30*(5), 551–64.

Jäger, Gerhard (2012). "Game theory in semantics and pragmatics." *Semantics, 3*, 2487–25.

Johnstone, Rufus A. (1995). "Sexual selection, honest advertisement and the handicap principle: reviewing the evidence." *Biological Reviews, 70*(1), 1–65.

Johnstone, Rufus A., and Alan Grafen (1992). "The continuous Sir Philip Sidney game: A simple model of biological signalling." *Journal of Theoretical Biology, 156*(2), 215–34.

Joyce, Richard (2007). *The evolution of morality*. Cambridge, MA: MIT Press.

Kandori, Michihiro, George J. Mailath, and Rafael Rob (1993). "Learning, mutation, and long run equilibria in games." *Econometrica: Journal of the Econometric Society*, *61*(1), 29–56.

Kane, Patrick, and Kevin J. S. Zollman (2015). "An evolutionary comparison of the handicap principle and hybrid equilibrium theories of signaling." *PloS One*, *10*(9), e0137271.

Keller, Laurent, and Kenneth G. Ross (1998). "Selfish genes: A green beard in the red fire ant." *Nature*, *394*(6693), 573–575.

Komarova, Natalia L., Partha Niyogi, Martin A. Nowak et al. (2001). "The evolutionary dynamics of grammar acquisition." *Journal of Theoretical Biology*, *209*(1), 43–59.

Lachmann, Michael, and Carl T. Bergstrom (2004). "The disadvantage of combinatorial communication." *Proceedings of the Royal Society of London B: Biological Sciences*, *271*(1555), 2337–43.

LaCroix, Travis, and Cailin O'Connor (2017). "Power by association." Unpublished.

Lehmann, Laurent, Katja Bargum, and Max Reuter (2006). "An evolutionary analysis of the relationship between spite and altruism." *Journal of Evolutionary Biology*, *19*(5), 1507–16.

Leimar, Olof, and Peter Hammerstein (2001). "Evolution of cooperation through indirect reciprocity." *Proceedings of the Royal Society of London B: Biological Sciences*, *268*(1468), 745–53.

Levins, Richard (1966). "The strategy of model building in population biology." *American Scientist*, *54*(4), 421–31.

Lewis, David (1969). *Convention: A philosophical study*. Cambridge, MA: Harvard University Press.

Lipman, Barton L. (2009). "Why is language vague?" Unpublished. http://people. bu. edu/blipman/Papers/vague5.pdf.

Martínez, Manolo (2015). "Deception in sender–receiver games." *Erkenntnis*, *80*(1), 215–27.

Martínez, Manolo (2019). Deception as cooperation. *Studies in History and Philosophy of Science Part C: Studies in History and Philosophy of Biological and Biomedical Sciences*, *77*, 101184. https://doi.org/10.1016/j.shpsc.2019.101184.

Martínez, Manolo, and Peter Godfrey-Smith (2016). "Common interest and signaling games: A dynamic analysis." *Philosophy of Science*, *83*(3), 371–92.

Martínez, Manolo, and Colin Klein (2016). "Pain signals are predominantly imperative." *Biology and Philosophy*, *31*(2), 283–98.

Martinez-Vaquero, Luis A., The Anh Han, Luís Moniz Pereira, and Tom Lenaerts (2015). "Apology and forgiveness evolve to resolve failures in cooperative agreements." *Scientific Reports*, 10639.

Maynard-Smith, John (1964). "Group selection and kin selection." *Nature*, *201*(4924), 1145–1147.

Maynard-Smith, John (1982). *Evolution and the theory of games*. Cambridge: Cambridge University Press.

Maynard-Smith, John (1991). "Honest signalling: The Philip Sidney game." *Animal Behaviour*, *42*(6), 1034–35.

Maynard-Smith, John. (2000). "The concept of information in biology." *Philosophy of Science*, *67*(2), 177–94.

Maynard-Smith, John, and David Harper (2003). *Animal signals*. Oxford: Oxford University Press.

Maynard-Smith, John, and G. R. Price (1973). "The logic of animal conflict." *Nature*, *246*(5247), 15–18.

McWhirter, Gregory (2015). "Behavioural deception and formal models of communication." *British Journal for the Philosophy of Science*, *67*(3), 757–80.

Millikan, Ruth Garrett (1984). *Language, thought, and other biological categories: New foundations for realism*. Cambridge, MA: MIT Press.

Millikan, Ruth Garrett (1995). "Pushmi-pullyu representations." *Philosophical Perspectives*, *9*, 185–200.

Mühlenbernd, Roland, and Michael Franke (2012). "Signaling conventions: Who learns what where and when in a social network?" In *The evolution of language*, 242–49. Singapore: World Scientific.

Nash, John (1950). "The bargaining problem." *Econometrica: Journal of the Econometric Society*, *18*(2), 155–162.

Nash, John (1953). "Two person cooperative games." *Econometrica: Journal of the Econometric Society*, *21*, 128–40.

Nowak, Martin A. (2006a). *Evolutionary dynamics*. Cambridge, MA: Harvard University Press.

Nowak, Martin A. (2006b). "Five rules for the evolution of cooperation." *Science*, *314*(5805), 1560–63.

Nowak, Martin A., and David C. Krakauer (1999). "The evolution of language." *Proceedings of the National Academy of Sciences*, *96*(14), 8028–33.

Nowak, Martin A., and Robert M. May (1992). "Evolutionary games and spatial chaos." *Nature*, *359*(6398), 828–829.

Nowak, Martin A., and Robert M. May (1993). "The spatial dilemmas of evolution." *International Journal of Bifurcation and Chaos*, *3*(1), 35–78.

Nowak, Martin A., and Karl Sigmund (1998). "Evolution of indirect reciprocity by image scoring." *Nature, 393*(6685), 573–577.

Nowak, Martin A., and Karl Sigmund (2005). "Evolution of indirect reciprocity." *Nature, 437*(7063), 1291–1298.

Nydegger, Rudy V., and Guillermo Owen (1974). "Two-person bargaining: An experimental test of the Nash axioms." *International Journal of Game Theory, 3*(4), 239–49.

O'Connor, Cailin (2014a). "The evolution of vagueness." *Erkenntnis, 79*(4), 707–27.

O'Connor, Cailin (2014b). "Evolving perceptual categories." *Philosophy of Science, 81*(5), 840–51.

O'Connor, Cailin (2015). "Ambiguity is kinda good sometimes." *Philosophy of Science, 82*(1), 110–21.

O'Connor, Cailin (2016). "The evolution of guilt: A model-based approach." *Philosophy of Science, 83*(5), 897–908.

O'Connor, Cailin (2017). "The cultural red king effect." *Journal of Mathematical Sociology, 41*(3), 155–171.

O'Connor, Cailin (2017). "Evolving to generalize: Trading precision for speed." *British Journal for the Philosophy of Science, 68*(2), 389–410.

O'Connor, Cailin (2019). *The origins of unfairness.* Oxford: Oxford University Press.

O'Connor, Cailin, Liam K. Bright, and Justin P. Bruner (2019). "The emergence of intersectional disadvantage." *Social Epistemology 33*(1), 23–41.

O'Connor, C., & Bruner, J. (2019). Dynamics and diversity in epistemic communities. *Erkenntnis, 84*(1), 101–119.

O'Connor, Cailin, and James Owen Weatherall (2016). "Black holes, black-scholes, and prairie voles: An essay review of simulation and similarity, by Michael Weisberg." *Philosophy of Science, 83*(4), 613–26.

Okamoto, Kyoko, and Shuichi Matsumura (2000). "The evolution of punishment and apology: An iterated prisoner's dilemma model." *Evolutionary Ecology, 14*(8), 703–20.

Okasha, Samir (2006). *Evolution and the levels of selection.* Oxford: Oxford University Press.

Okasha, Samir (2013). "Biological altruism." In *The Stanford encyclopedia of philosophy.* Ed. Edward N. Zalta. https://plato.stanford.edu/entries/altruism-biological/.

Okasha, Samir, and Johannes Martens (2016). "Hamilton's rule, inclusive fitness maximization, and the goal of individual behaviour in symmetric two-player games." *Journal of Evolutionary Biology, 29*(3), 473–82.

Panchanathan, Karthik, and Robert Boyd (2004). "Indirect reciprocity can stabilize cooperation without the second-order free rider problem." *Nature, 432*(7016), 499–502.

Pawlowitsch, Christina (2007). "Finite populations choose an optimal language." *Journal of Theoretical Biology, 249*(3), 606–16.

Pawlowitsch, Christina, et al. (2008). "Why evolution does not always lead to an optimal signaling system." *Games and Economic Behavior, 63*(1), 203–26.

Planer, Ronald J. (2014). "Replacement of the 'genetic program' program." *Biology and Philosophy, 29*(1), 33–53.

Poza, David J., Félix A. Villafáñez, Javier Pajares, Adolfo López-Paredes, and Cesáreo Hernández (2011). "New insights on the Emergence of Classes Model." *Discrete Dynamics in Nature and Society, 2011,* 915279.

Price, George R., et al. (1970). "Selection and covariance." *Nature, 227,* 520–21.

Quine, Willard V. (1976). "Truth by convention." In *The ways of paradox, and other essays.* 1936, repr. Cambridge, MA: Harvard University Press.

Robert Aumann, J.J Gabszewski, JF Richard, LA Wolsey (1990). "Nash equilibria are not self-enforcing." *Economic Decision Making: Games, Econometrics and Optimisation,* 201–6.

Robinson, Joan (1962). *Economic philosophy.* Middlesex, UK: Penguin Books.

Robinson, Julia (1951). "An iterative method of solving a game." *Annals of Mathematics, 54*(2), 296–301.

Robson, Arthur J. (1990). "Efficiency in evolutionary games: Darwin, Nash and the secret handshake." *Journal of Theoretical Biology, 144*(3), 379–96.

Rosenstock, Sarita, and Cailin O'Connor (2018). "When it's good to feel bad: An evolutionary model of guilt and apology." *Frontiers in Robotics and AI, 5,* 9. https://doi.org/10.3389/frobt.2018.00009.

Roth, Alvin E., and Ido Erev (1995). "Learning in extensive-form games: Experimental data and simple dynamic models in the intermediate term." *Games and Economic Behavior, 8*(1), 164–212.

Rousseau, Jean-Jacques (1984). *A discourse on inequality.* New York: Penguin.

Rubin, Hannah (2016). "The phenotypic gambit: selective pressures and ESS methodology in evolutionary game theory." *Biology and Philosophy, 31*(4), 551–69.

Rubin, Hannah (2018). "The debate over inclusive fitness as a debate over methodologies." *Philosophy of Science, 85*(1), 1–30.

Rubin, Hannah, Justin Bruner, Cailin O'Connor, and Simon Huttegger (2015). "Communication without common interest: A signaling experiment." Unpublished.

Rubin, Hannah, and Cailin O'Connor (2018). "Discrimination and collaboration in science." *Philosophy of Science*, 85(3), 380–402.

Rubinstein, Ariel (1982). "Perfect equilibrium in a bargaining model." *Econometrica: Journal of the Econometric Society*, 50(1), 97–109.

Santana, Carlos (2014). "Ambiguity in cooperative signaling." *Philosophy of Science*, 81(3), 398–422.

Santos, Francisco C., Jorge M. Pacheco, and Brian Skyrms (2011). "Co-evolution of pre-play signaling and cooperation." *Journal of Theoretical Biology*, 274(1), 30–35.

Schupbach, Jonah N. (2016). "Robustness analysis as explanatory reasoning." *British Journal for the Philosophy of Science*, 69(1), 275–300.

Searcy, William A., and Stephen Nowicki (2005). *The evolution of animal communication: Reliability and deception in signaling systems*. Princeton, NJ: Princeton University Press.

Seyfarth, Robert M., Dorothy L. Cheney, and Peter Marler (1980a). "Monkey responses to three different alarm calls: Evidence of predator classification and semantic communication." *Science*, 210(4471), 801–3.

Seyfarth, Robert M., Dorothy L. Cheney, and Peter Marler (1980b). "Vervet monkey alarm calls: semantic communication in a free-ranging primate." *Animal Behaviour*, 28(4), 1070–94.

Shannon, Claude Elwood (1948). "A mathematical theory of communication." *Bell System Technical Journal*, 27(3), 379–423.

Shea, Nicholas (2007). "Representation in the genome and in other inheritance systems." *Biology and Philosophy*, 22(3), 313–31.

Shea, Nicholas (2011). "What's transmitted? Inherited information." *Biology and Philosophy*, 26(2), 183–89.

Shea, Nicholas (2013). "Inherited representations are read in development." *British Journal for the Philosophy of Science*, 64(1), 1–31.

Shea, Nicolas, Peter Godfrey-Smith, and Rosa Cao (2018). "Content in simple signalling systems." *British Journal for the Philosophy of Science*, 69(4), 1009–35.

Sigmund, Karl (2011). "Introduction to evolutionary game theory." *Evolutionary Game Dynamics*, 69, 1–26.

Sigmund, Karl (2017). *Games of life: Explorations in ecology, evolution and behavior*. New York: Courier Dover.

Silk, Joan B., Elizabeth Kaldor, and Robert Boyd (2000). "Cheap talk when interests conflict." *Animal Behaviour*, 59(2), 423–32.

Skyrms, Brian (1996). *Evolution of the social contract*. Cambridge: Cambridge University Press.

Skyrms, Brian (2002a). "Altruism, inclusive fitness, and 'the logic of decision.'" *Philosophy of Science, 69*(S3), S104–11.

Skyrms, Brian (2002b). "Signals, evolution and the explanatory power of transient information." *Philosophy of Science, 69*(3), 407–28.

Skyrms, Brian (2004). *The stag hunt and the evolution of social structure.* Cambridge: Cambridge University Press.

Skyrms, Brian (2009). "Evolution of signalling systems with multiple senders and receivers." *Philosophical Transactions of the Royal Society B: Biological Sciences, 364*(1518), 771–79.

Skyrms, Brian (2010a). "The flow of information in signaling games." *Philosophical Studies, 147*(155), https://doi.org/10.1007/s11098-009-9452-0

Skyrms, Brian (2010b). *Signals: Evolution, learning, and information.* Oxford: Oxford University Press.

Skyrms, Brian (2012). "Learning to signal with probe and adjust." *Episteme, 9*(2), 139–50.

Skyrms, Brian, and Jeffrey Barrett (2018). "Propositional content in signals." Unpublished.

Skyrms, Brian, and Robin Pemantle (2000). "A dynamic model of social network formation." *Proceedings of the National Academy of Sciences, 97*(16), 9340–46.

Skyrms, Brian, and Kevin J. S. Zollman (2010). "Evolutionary considerations in the framing of social norms." *Politics, Philosophy, and Economics, 9*(3), 265–73.

Smead, Rory (2010). "Indirect reciprocity and the evolution of 'moral signals.'" *Biology and Philosophy, 25*(1), 33–51.

Smead, Rory (2012). "Game theoretic equilibria and the evolution of learning." *Journal of Experimental and Theoretical Artificial Intelligence, 24*(3), 301–13.

Smead, Rory (2014a). "Deception and the evolution of plasticity." *Philosophy of Science, 81*(5), 852–65.

Smead, Rory (2014b). "Evolving games and the social contract." In *Complexity and the human experience*, 61–80. Ed. Paul A. Youngman and Mirsad Hadzikadic. New York: Jenny Stanford.

Smead, Rory (2014c). "The role of social interaction in the evolution of learning." *British Journal for the Philosophy of Science, 66*(1), 161–80.

Smead, Rory, and Patrick Forber (2013). "The evolutionary dynamics of spite in finite populations." *Evolution, 67*(3), 698–707.

Smead, Rory, and Patrick Forber (2016). "The coevolution of recognition and social behavior." *Scientific Reports, 6*, 25813.

Smead, Rory, and Kevin J. S. Zollman (2009). "The stability of strategic plasticity." Carnegie Mellon Technical Report.

Smith, Vernon L. (1994). "Economics in the Laboratory." *Journal of Economic Perspectives, 8*(1), 113–31.

Sober, Elliott, and David Sloan Wilson (1999). *Unto others: The evolution and psychology of unselfish behavior.* Cambridge, MA: Harvard University Press.

Spence, Michael (1978). "Job market signaling." In *Uncertainty in economics,* 281–306. New York: Elsevier.

Steinert-Threlkeld, Shane (2016). "Compositional signaling in a complex world." *Journal of Logic, Language and Information, 25*(3–4), 379–97.

Sterelny, Kim (2012). *The evolved apprentice.* Cambridge, MA: MIT Press.

Stewart, Quincy Thomas (2010). "Big bad racists, subtle prejudice and minority victims: An agent-based analysis of the dynamics of racial inequality." Paper presented at the annual meeting of the Population Association of America.

Taylor, Peter D., and Leo B. Jonker (1978). "Evolutionary stable strategies and game dynamics." *Mathematical Biosciences, 40*(1–2), 145–56.

Thorndike, Edward L. (1898). "Animal intelligence: An experimental study of the associative processes in animals." *Psychological Review: Monograph Supplements, 2*(4).

Tomasello, Michael, Alicia P. Melis, Claudio Tennie, Emily Wyman, Esther Herrmann, Ian C. Gilby, Kristen Hawkes, Kim Sterelny, Emily Wyman, Michael Tomasello et al. (2012). "Two key steps in the evolution of human cooperation: The interdependence hypothesis." *Current Anthropology, 53*(6), 673–692.

Trivers, Robert L. (1971). "The evolution of reciprocal altruism." *Quarterly Review of Biology, 46*(1), 35–57.

UW Health (2015). "Longest kidney chain ever completed" https://www.uwhealth.org/news/longest-kidney-chain-ever-completed-wraps-up-at-uw-hospital-and-clinics/45549

Van Rooij, Robert, and Michael Franke (2015). "Optimality-theoretic and game-theoretic approaches to implicature." *The Stanford Encyclopedia of Philosophy.* Ed. Edward N. Zalta. https://plato.stanford.edu/entries/implicature-optimality-games/

Vanderschraaf, Peter (1995). "Convention as correlated equilibrium." *Erkenntnis, 42*(1), 65–87.

Vanderschraaf, Peter (1998a). "The informal game theory in Hume's account of convention." *Economics and Philosophy, 14*(2), 215–47.

Vanderschraaf, Peter (1998b). "Knowledge, equilibrium and convention." *Erkenntnis, 49*(3), 337–69.

Vanderschraaf, Peter, and J. McKenzie Alexander (2005). "Follow the leader: Local interactions with influence neighborhoods." *Philosophy of Science, 72*(1), 86–113.

Ventura, Rafael (2019). "Ambiguous signals, partial beliefs, and propositional content." *Synthese, 196*(7), 2803–2820.

Von Neumann, John, and Oskar Morgenstern (1944). *Theory of games and economic behavior*. Princeton, NJ: Princeton University Press.

Wagner, Elliott (2009). "Communication and structured correlation." *Erkenntnis, 71*(3), 377–93.

Wagner, Elliott (2013a). "The explanatory relevance of Nash equilibrium: One-dimensional chaos in boundedly rational learning." *Philosophy of Science, 80*(5), 783–95.

Wagner, Elliott O. (2011). "Deterministic chaos and the evolution of meaning." *British Journal for the Philosophy of Science, 63*(3), 547–75.

Wagner, Elliott O. (2012). "Evolving to divide the fruits of cooperation." *Philosophy of Science, 79*(1), 81–94.

Wagner, Elliott O. (2013b). "The dynamics of costly signaling." *Games, 4*(2), 163–81.

Wärneryd, Karl (1993). "Cheap talk, coordination, and evolutionary stability." *Games and Economic Behavior, 5*(4), 532–46.

Weibull, Jörgen W. (1997). *Evolutionary game theory*. Cambridge, MA: MIT Press.

Weisberg, Michael (2012). *Simulation and similarity: Using models to understand the world*. Oxford: Oxford University Press.

West, Stuart A., Claire El Mouden, and Andy Gardner (2011). "Sixteen common misconceptions about the evolution of cooperation in humans." *Evolution and Human Behavior, 32*(4), 231–62.

West, Stuart A., and Andy Gardner (2010). "Altruism, spite, and greenbeards." *Science, 327*(5971), 1341–44.

Wilson, David Sloan, and Elliott Sober (1994). "Reintroducing group selection to the human behavioral sciences." *Behavioral and Brain Sciences, 17*(4), 585–608.

Yaari, Menahem E., and Maya Bar-Hillel (1984). "On dividing justly." *Social Choice and Welfare, 1*(1), 1–24.

Young, H. Peyton (1993). "An evolutionary model of bargaining." *Journal of Economic Theory, 59*(1), 145–68.

Zahavi, Amotz (1975). "Mate selection—a selection for a handicap." *Journal of Theoretical Biology, 53*(1), 205–14.

Zimen, Erik (1981). *The wolf: His place in the natural world*. London: Souvenir Press.

Zollman, Kevin J. S. (2005). "Talking to neighbors: The evolution of regional meaning." *Philosophy of Science, 72*(1), 69–85.

Zollman, Kevin J. S. (2011). "Separating directives and assertions using simple signaling games." *Journal of Philosophy, 108*(3), 158–69.

Zollman, Kevin J. S. (2013). "Finding alternatives to handicap theory." *Biological Theory, 8*(2), 127–32.

Zollman, Kevin J. S., Carl T. Bergstrom, and Simon M. Huttegger (2012). "Between cheap and costly signals: The evolution of partially honest communication." *Proceedings of the Royal Society of London B: Biological Sciences* https://doi.org/10.1098/rspb.2012.1878, *280*, 20121878.

Zollman, Kevin J. S., and Rory Smead (2010). "Plasticity and language: An example of the Baldwin effect?" *Philosophical Studies, 147*(7), 7. https://doi.org/10.1007/s11098-009-9447-x.

Acknowledgments

Many thanks to those who were willing to share feedback, and reading suggestions, during the preparation of this Element. This list includes Jason Alexander, Jeffrey Barrett, Jonathan Birch, Justin Bruner, Simon Huttegger, Travis LaCroix, Rory Smead, Hannah Rubin, Nicholas Shea, and Brian Skyrms. My thanks to Grant Ramsey and Michael Ruse for organizing this Elements series, and for suggestions on the initial topic proposal. Thanks to the editorial staff at Cambridge for feedback and work on this manuscript. And thanks, as always to Jim Weatherall for moral support. And to Eve and Vera for helping improve my work/life balance.

Cambridge Elements $\equiv$

Philosophy of Biology

Grant Ramsey
KU Leuven

Grant Ramsey is a BOFZAP Research Professor at the Institute of Philosophy, KU Leuven, Belgium. His work centers on philosophical problems at the foundation of evolutionary biology. He has been awarded the Popper Prize twice for his work in this area. He also publishes in the philosophy of animal behavior, human nature, and the moral emotions. He runs the Ramsey Lab (theramseylab.org), a highly collaborative research group focused on issues in the philosophy of the life sciences.

Michael Ruse
Florida State University

Michael Ruse is the Lucyle T. Werkmeister Professor of Philosophy and the Director of the Program in the History and Philosophy of Science at Florida State University. He is Professor Emeritus at the University of Guelph, in Ontario, Canada. He is a former Guggenheim fellow and Gifford lecturer. He is the author or editor of over sixty books, most recently *Darwinism as Religion: What Literature Tells Us about Evolution; On Purpose; The Problem of War: Darwinism, Christianity, and their Battle to Understand Human Conflict;* and *A Meaning to Life.*

About the Series

This Cambridge Elements series provides concise and structured introductions to all of the central topics in the philosophy of biology. Contributors to the series are cutting-edge researchers who offer balanced, comprehensive coverage of multiple perspectives, while also developing new ideas and arguments from a unique viewpoint.

Cambridge Elements ⁼

Philosophy of Biology

CPSIA information can be obtained
at www.ICGtesting.com
Printed in the USA
LVHW051936040220
645828LV00020B/864